Frederick A. P. Barnard

Metric System of Weights and Measures

Frederick A. P. Barnard

Metric System of Weights and Measures

ISBN/EAN: 9783337192396

Printed in Europe, USA, Canada, Australia, Japan

Cover: Foto ©berggeist007 / pixelio.de

More available books at **www.hansebooks.com**

THE

METRIC SYSTEM

OF

WEIGHTS AND MEASURES.

THE

METRIC SYSTEM

OF

WEIGHTS AND MEASURES;

AN ADDRESS DELIVERED BEFORE THE CONVOCATION OF THE
UNIVERSITY OF THE STATE OF NEW YORK,
AT ALBANY, AUGUST 1, 1871 ;

BY FREDERICK A. P. BARNARD, S. T. D., LL. D.,

President of Columbia College, New York City ;

MEMBER OF THE NATIONAL ACADEMY OF SCIENCES; AND OF THE AMERICAN PHILOSOPHICAL SOCIETY,
PHILADELPHIA; ASSOCIATE MEMBER OF THE AMERICAN ACADEMY OF ARTS AND SCIENCES,
BOSTON; CORRESPONDING MEMBER OF THE ROYAL SOCIETY OF
SCIENCES OF LIÈGE, BELGIUM, &C., &C.

———

REVISED EDITION.

PRINTED BY ORDER OF THE BOARD OF TRUSTEES OF COLUMBIA COLLEGE.

———

NEW YORK
1872.

CONTENTS.

PREFACE.

On the eighth day of August, 1866, the Convocation of the University of the State of New York, being then in session at Albany, was addressed by the Hon. JOHN A. KASSON, a member of the House of Representatives of the Congress of the United States from Iowa, and Chairman of a Committee of that body appointed to consider the possibility of securing a uniform system of Coinage, Weights and Measures for all nations. Congress had then recently (July 27th, 1866), passed an act legalizing the use of the metrological system known as "The Metric System of Weights and Measures," in all business transactions in the United States; and, two years earlier, a similar act had passed the British Parliament. It was known that the system had met with large acceptance on the continent of Europe, and also in the greater portion of the American continent south of our own territory; and it was also known that the use of this system was becoming more extended every year.

The aim and hope of Mr. KASSON had been that he might enlist the large body of enlightened educators forming the University Convocation, in an active effort to advance the cause of metrological reform in our country, by diffusing among the people information in regard to the Metric System; by pointing out the merits of this system; and by meeting the objections with which the proposition to naturalize it here, like every other threatened innovation upon established usages, however in its own nature desirable or prospectively beneficent, is sure to be encountered. And the appearances at the time were certainly favorable to the fulfilment of this hope; for the address of the honorable gentleman was received with evident marks of approval.

A committee was accordingly appointed, charged with the duty of reporting on the subject to the Convocation at a future meeting. This committee consisted of the Hon. JOHN V. L. PRUYN, LL.D. (Chancellor of the University), CHARLES DAVIES, LL.D., Professor Emeritus of the Higher Mathematics in Columbia College, and ROBERT S. HALE, LL.D., one of the Regents of the University. Prof. DAVIES, who was charged with the preparation of the report, states that it originally "seemed to be the unanimous opinion of the committee that a report would be made favorable to the introduction of the [metric] system into general use;" but that reflection and inquiry led to a modification of views, especially on his own part; and that the conclusion was reached that the Convocation should not "commit itself hastily to the great and radical changes which the introduction of the metric system would occasion." It was not, therefore, until after three years of deliberation, that the committee presented their report; and the report then made, which is said to have been partial, was apparently oral.

This committee was thereupon discharged, and a new one appointed, consisting of Prof. DAVIES, Regent HALE, and Prof. JAMES B. THOMSON, LL.D. It is stated in the preface to the final report, that Prof. THOMSON did not act with the committee. The report of this reconstructed committee was presented to the Convocation at the session of August, 1870. It consisted mainly in an argument to demonstrate the inexpediency and impracticability of introducing the Metric

System of Weights and Measures into the United States. By order of the Convocation this report was published and extensively circulated.

The Trustees and Faculty of the College with which the chairman of the Committee held formerly an official, and holds still an honorary, connection, have for some years been upon the record as advocates of legislation by the Congress of the United States, favorable to the unification of the Moneys, Weights and Measures of the world. In their view, the object desired, so far as it regards Weights and Measures, is most likely to be secured through the universal acceptance of a metrological system which is already generally received; and that is the Metric System. To them it appeared that the publication of a report prepared by a gentleman in nominal connection with them, maintaining an opposite opinion, was likely to produce an erroneous impression in the public mind in regard to their own position. At a meeting therefore of the Trustees, held on the first day of May, 1871, a resolution was adopted, on motion of the Hon. SAMUEL BLATCHFORD, LL.D., Judge of the United States District Court for the Southern District of New York, requesting the President of the College to attend the meeting of the Convocation to be held in the August next ensuing, and to state to that body how far the views set forth in the report of the committee above referred to are in harmony with those entertained by the Faculty of the College.

It was in obedience to this resolution that the address contained in the following pages was prepared. The address was listened to with evidently interested attention by the Convocation; and, by the courtesy of the Regents of the University, it was immediately published, in advance of the report of the Proceedings of the Convocation, in pamphlet form. Some copies of this publication having been laid before the

Trustees of Columbia College at a meeting held on the second day of October, 1871, it was, on motion of Judge BLATCHFORD, resolved, that a revised edition of one thousand copies be printed for general circulation. In the present edition, issued in conformity with this order, some slight modifications have been made in the original text; and some additional information presumedly of interest has been appended in the form of notes.

ORIGIN AND NATURE

OF THE

METRIC SYSTEM OF WEIGHTS AND MEASURES.

No cause, since the earliest organization of civilized society, has contributed more largely to embarrass business transactions among men, especially by interfering with the facility of commercial exchanges between different countries, or between different provinces, cities, or even individual citizens of the same country, than the endless diversity of instrumentalities employed for the purpose of determining the *quantities* of exchangeable commodities. For the inconvenience and confusion resulting from this cause, but one effectual remedy can possibly be suggested; and that is the general adoption throughout the world of one common system of weights and measures. Until nearly the close of the eighteenth century, nevertheless, no movement appears to have been anywhere made, looking to the immediate or prospective application of this remedy. It was one of the projects entertained by the Constituent Assembly of France, at a time when the revolution had not yet passed into that sanguinary phase which but too soon succeeded, to engage the nations of Europe in a united effort to create, for the common use of all, a new metrological system, founded upon standards determined with scientific accuracy, and constructed in its details according to a scientific method. Nor, amid all the succeeding excitements attendant on the downfall of the monarchy, and the inauguration of the republic and the " Terror," was this important object ever lost sight of by the men who held successively in their hands the destinies of France. And though the convulsions which, for many successive years during that stormy period, agitated the continent of Europe, prevented the participation of all the nations in the prosecution of this great and beneficent work, still the work itself was prosecuted, though with some interruptions, to a satisfactory completion; and the result is seen to day in the Metric System of Weights and Measures; a system which, after the lapse of only three quarters of a century, has been adopted for use by more than half the inhabitants of the civilized and Christian world.

The principles according to which this system has been constructed are set forth in the following statement, adopted from the report on the subject made by the Honorable JOHN QUINCY ADAMS, Secretary of State of the United States, to the House of Representatives of the Sixteenth Congress, under date of February 22d, 1821.

1. That all weights and measures should be reduced to one *uniform* standard of linear measure.

2. That this standard should be an aliquot part of the circumference of the globe.

3. That the unit of linear measure, applied to matter in its three modes of extension, length, breadth, and thickness, should be the standard of all measures of length, surface, and solidity.

4. That the cubic contents of the linear measure, in distilled water, at the temperature of its greatest contraction,

should furnish at once the standard weight and measure of capacity.

5. That for everything susceptible of being measured or weighed, there should be only one measure of length, one weight, one measure of contents, with their multiples and subdivisions exclusively in decimal proportions.

6. That the principle of decimal division, and a proportion to the linear standard, should be annexed to the coins of gold, silver, and copper, to the moneys of account, to the division of time, to the barometer and thermometer, to the plummet and log-lines of the sea, to the geography of the earth and the astronomy of the skies; and, finally, to everything in human existence susceptible of comparative estimation by weight or measure.

7. That the whole system should be equally suitable to the use of all mankind.

8. That every weight and every measure should be designated by an appropriate, significant, characteristic name, applied exclusively to itself.

The following is the succinct account given by Mr. ADAMS, of the early history of the movement in which the Metric System had its origin:

"In the year 1790, the Prince de Talleyrand, then Bishop of Autun, distributed among the members of the Constituent Assembly of France, a proposal, founded upon the excessive diversity and confusion of the weights and measures then prevailing all over that country, for the reformation of the system, or, rather, for the foundation of a new one, upon the principle of a single and universal standard. After referring to the two objects which had previously been suggested by Huyghens and Picard—the pendulum, and the proportional part of the circumference of the earth—he concluded by giving the preference to the former, and presented the project of a decree. First, that exact copies of all the different weights and elementary measures *used* in every town of France, should be obtained and sent to Paris. Secondly, that the National Assembly should write a letter to the British Parliament, requesting their concurrence with France in the adoption of a natural standard for weights and measures; for which purpose, Commissioners, in equal numbers from the French Academy of Sciences, and the British Royal Society, chosen by those learned bodies respectively, should meet at the most suitable place, and ascertain the length of the pendulum at the forty-fifth degree of latitude, and from it an invariable standard for all measures and weights. Thirdly, that, after the accomplishment, with all due solemnity, of this operation, the French Academy of Sciences should fix with precision the tables of proportion between the new standards, and the weights and measures previously used in the various parts of France; and that every town should be supplied with exact copies of the new standards, and with tables of comparison between them and those of which they were to supply the place. This decree, somewhat modified, was adopted by the Assembly; and, on the 22d of August, 1790, sanctioned by Louis the Sixteenth. Instead of writing to the British Parliament themselves, the Assembly requested the King to write to the King of Great Britain, inviting him to propose to the Parliament the formation of a joint commission of members of the Royal Society and of the Academy of Sciences, to ascertain the natural standard in the length of the pendulum. Whether the forms of the British Constitution, the temper of political animosity then subsisting between the two countries, or the convulsions and wars which soon afterwards ensued, prevented the acceptance and execution of this proposal, it is deeply to be lamented that it was not carried into effect.

*   *   *   *   *

"The idea of associating the interests and the learning of other nations in this great effort for common improvement was not confined to the proposal for obtaining the concurrent agency of Great Britain. Spain, Italy, the Netherlands, Denmark, and Switzerland were actually represented in the proceedings of the Academy of Sciences to accomplish the purposes of the National Assembly. But, in the first instance, a committee of the Academy of Sciences, consisting of five of the ablest members of the Academy and most eminent mathematicians of Europe, Borda, Lagrange, Laplace, Monge, and Condorcet, were chosen, under the decree of the assembly, to report to that body upon the selection of the natural standard and other principles proper for the accomplishment of the object. Their report to the Academy was made on the 19th of March, 1791, and immediately transmitted to the national assembly, by whose orders it was printed. The committee, after examining three projects of a natural standard, the pendulum beating seconds, a quarter of the equator, and a quarter of the meridian, had, on a full deliberation, and with great accuracy of judgment, preferred the last, and proposed that its ten-millionth part should be taken as the standard unit of linear measure; that, as a second standard of comparison with it, the pendulum vibrating seconds at the forty-fifth degree of latitude should be assumed, and that the weight of distilled water at the point of freezing, measured by a cubical vessel in decimal proportion to the linear standard, should determine the standard of weights and of vessels of capacity."

This report having received the sanction of the assembly,

committees of the Academy of Sciences were appointed to make with all the necessary precision the determinations upon which were to rest the standard units of the new metrological system. The most laborious of these determinations consisted in the trigonometrical measurement of an arc of the meridian extending through France from Dunkirk to Barcelona; an operation which occupied seven years. The design of this was to determine with exactness the length of the linear base, called the METRE. But the assembly did not wait for the completion of this great work before giving to the system a legal and a practical existence; the length of the degree of latitude being already known with a sufficiently near approach to exactness to make any possible error of the metre founded upon it entirely insensible for the ordinary purposes of life. The system was therefore provisionally established by a law of the 1st of August, 1793; and the nomenclature which now distinguishes it was adopted on the 18th Germinal, An. III., (7th April, 1795).

In the seventh year of the Republic (1799) an international commission was assembled at Paris, on the invitation of the government, to settle, from the results of the great Meridian Survey, the exact length of the " definitive metre." In this commission were represented the governments of France, Holland, Denmark, Sweden, Switzerland, Spain, Savoy, and the Roman, Cisalpine and Ligurian Republics. After the completion of its labors, the commission proceeded on the 4th Messidor, An. VII. (22d June, 1799), to deposit, at the Palace of the Archives, in Paris, the standard metre-bar of platinum, which represents the linear base of the system; and the standard kilogramme weight, also of platinum, which represents the unit of metric weights.

Of these prototype standards, numerous copies have been

taken, which, after having been compared with the originals with the severest exactness, have been made standards of reference and verification in the various countries in which the system has been adopted.

THE METRIC SYSTEM, founded on the metre as the unit of length, has four other leading units, all connected with and dependent upon this.   Hence we have—

1. The METRE, which is the unit of measures of length.

2. The ARE, which is the unit of measures of surface, and is the square of ten metres.

3. The LITRE, which is the unit of measures of capacity, and is the cube of a tenth part of the metre.

4. The STERE, which is the unit of measures of solidity, having the capacity of a cubic metre.

5. The GRAMME, which is the unit of measures of weight, and is the weight of that quantity of distilled water, at its maximum density, which fills the cube of the hundredth part of the metre.

Each unit has its decimal multiples and submultiples, that is, weights and measures ten times larger or ten times smaller than the principal unit.   These multiples and submultiples are indicated by prefixes placed before the names of the several fundamental units.   The prefixes denoting multiples are derived from the Greek language; and are *deka*, ten; *hecto*, hundred; *kilo*, thousand; and *myria*, ten thousand. Those denoting submultiples are taken from the Latin; and are, *deci*, tenth; *centi*, hundredth; and *milli*, thousandth.

The following table embraces all the weights and measures of the system:

RELATIVE VALUE.	LENGTH.	SURFACE.	CAPACITY.	SOLIDITY.	WEIGHT.
10,000....	Myria-metre...				
1,000....	Kilo-metre..............		Kilo-litre..............		Kilo-gramme.
100....	Hecto-metre..	Hect-are....	Hecto-litre............		Hecto-gramme.
10....	Deka-metre............		Deka-litre...	Deka-stere.	Deka-gramme.
UNIT.	METRE.	ARE.	LITRE.	STERE.	GRAMME.
.1...	Deci-metre..	Deci-are....	Deci-litre....	Deci-stere..	Deci-gramme.
.01..	Centi-metre..	Centi-are...	Centi-litre.............		Centi-gramme.
.001.	Milli-metre.............		Milli-litre..............		Milli-gramme.

The denominations of solid measure beyond the first multiple and sub-multiple by *ten* are not in use. The term *stere* itself is in fact rarely employed, measures of solidity or volume being usually expressed in cubic denominations of the linear base. Of agrarian measures, the only derivatives of the unit in use are the hectare, the deciare, and the centi-are.

VALUES OF UNITS.

UNIT OF LENGTH.

Myriametre..........	10,000	Metres.
Kilometre...........	1,000	"
Hectometre..........	100	"
Decametre..........	10	"
METRE..............	1	"
Decimetre..........	0.1	" or $\frac{1}{10}$ of a metre.
Centimetre..........	0.01	" " $\frac{1}{100}$ " "
Millimetre..........	0.001	" " $\frac{1}{1000}$ " "

The METRE is equal to 3.280899 feet nearly ; or to 39.37079 inches.

The unit of Itinerary measure is the KILOMETRE, which is equal to 0.62138 miles.

UNIT OF SURFACE.

Hectare....	100	Ares or 1000 Square Metres.				
ARE.......	1	"	"	100	"	"
Deciare....	0.1	"	"	10	"	"
Centiare....	0.01	"	"	1	"	"

The ARE is equal to 119.60332 square yards; the HECTARE, the Agrarian unit, to 2.47114 acres.

UNIT OF CAPACITY.

Hectolitre	=	100	Litres.			
Dekalitre	=	10	"			
LITRE	=	1	"			
Decilitre	=	0.1	"	or $\frac{1}{10}$ of a Litre.		
Centilitre	=	0.01	"	" $\frac{1}{100}$	"	"
Millilitre	=	0.001	"	" $\frac{1}{1000}$	"	"

The LITRE is equal to 0.26418635 gallons, or 1.0567454 quarts, or 2.1134908 pints.

UNIT OF WEIGHT.

Kilogramme	1000	Grammes.				
Hectogramme	100	"				
Dekagramme	10	"				
GRAMME	1	"				
Decigramme	0.1	"	or $\frac{1}{10}$ of a gramme.			
Centigramme	0.01	"	" $\frac{1}{100}$	"	"	
Milligramme	0.001	"	" $\frac{1}{1000}$	"	"	

The GRAMME is equal to 15.43234874 grains.*

The KILOGRAMME, which is the unit of commercial weight, is equal to 2.20462125 pounds avoirdupois.

---

* This is the Gramme as derived from the weight *in vacuo* of the platinum Kilogramme of the Archives, as determined by Prof. MILLER of London, in 1844, and adopted by the Standards-Department of the British Government. See APPENDIX B.

THE METRIC SYSTEM

OF

WEIGHTS AND MEASURES.

———·•·———

I.—*Recent Progress of Metrological Reform.*

INTRODUCTION.

GENTLEMEN OF THE CONVOCATION :

The sense of the right of property is an instinctive feeling, of which the existence is co-extensive with intelligence. We find abundant evidence of its presence in the lower animals as well as in ourselves. The dog, for instance, when he has satisfied his hunger, carefully stores up the superfluous bone of to-day, in prudent provision for the anticipated wants of the morrow. The beast of the forest bears his prey to the lair which he has appropriated to himself; and the birds defend with spirit the nests which their own labors have constructed. In the social animals, as the beaver, and the social insects, as the ant and the bee, we see the principle more broadly developed. In these cases, the dwelling which the common toil has constructed or prepared, and the stores which the common industry has gathered, are the common property of all; and are apportioned for the benefit of individuals upon principles which we probably do not understand. But the lower animals, though they appropriate to themselves articles which seem desirable, and assert a right of property in the objects thus appropriated, never propose to relinquish one possession in consideration of an equivalent offered in the form of another. They have no notion of commerce or exchange even in its simplest form. The commercial idea makes its first appearance in man. It is present in every stage of human civilization. Its earliest practical illustration is in the form of barter, in which objects supposed to have value are exchanged one against the other; or a single one of a certain description for several of another. But as wealth increases, and as its forms become more diversified, the necessity of determining equivalents by quantity rather than by tale, becomes manifest; and out of this necessity springs the creation of conventional standards, by means of which quantities may be always and everywhere verified, and definite quantities be correctly ascertained. Hence have arisen the various systems of weight and measure which have prevailed among different peoples, some form of which has been found to accompany even the rudest civilization. Such systems having originated before anything like intellectual culture existed, have been constructed without any thought of scientific method, and have owed their earliest form to accident or caprice. As social and political institutions have become more fully developed, legislation has stepped in from time to time to alter if not to improve these primitive systems; to change the value of their unit bases ; or to modify the relations to these bases of the derivative denominations ; until, at the present time, there is no reason to believe that there survives in any existing system of weights and measures a single value of any unit identical with one in use two thousand years ago ; or a law of derivation connecting the different branches of the system, the weights and the measures of capacity, for instance, with the linear base, such as governed the same relations at a period so far remote in the past as the earlier ages of the christian era.

To change systems of weight and measure, and to change them by legislation, is therefore no new thing, first thought of in our day. It is a thing which has been going on ever since the birth of civilization. It is not in itself a good thing, or a desirable thing, or a thing we should engage in for its own sake ; neither is it a desirable thing to pull down the dwelling in which we have long lived, even though it may be inconvenient. Time has, perhaps, reconciled us to the inconveniences ; use has made us forgetful of their existence ; or habit has possibly, for such is human nature, converted defects into merits in our eyes. But when

we do pull down the old home, and we often do, and
Americans do so probably oftener than they ought,
and in its place erect an edifice constructed on better
principles, we find, in the increased comfort which follows, a justification of the proceeding, and a compen-
sation for all the trouble and expense it has cost us.

There has grown up within our own century a
branch of systematic inquiry, which in recent years
has been prosecuted with a high degree of activity,
under the name of "social science." This inquiry com-
prehends in its scope every problem interesting to
human society, political, educational, moral, econom-
ical, commercial, statistical, and sanitary ; its aim
being not merely the enlargement of knowledge for
its own sake, but the practical amelioration of the
condition of man. In an investigation so comprehensive, the subject of weights, measures and coins, the
instruments by means of which the values of all
objects which make up the wealth of nations must be
measured, and by means of which also the exchanges
of commodities which constitute the world's commerce
must be effected, could not escape attention. An
earnest movement has accordingly been going on in
our time which, during the past twenty years, has
been pressed with especial urgency, having for its
object to remove the impediments to freedom and
facility of commercial intercourse between nations, and
the obstacles to the intelligent understanding, on the
part of individuals, of the material condition of the

world and the progress of contemporaneous history, which arise from the diversity and discordance of existing metrological systems. At a former meeting of this convocation, an ardent friend of the movement here spoken of, a gentleman who was at that time a member of our national legislature, addressed you on this subject, and recommended it earnestly to your favorable consideration. The result was disappointing. The movement failed to command from you the anticipated sympathy ; and in approaching the same subject to-day, I feel how heavy is the task of one who attempts to plead before you a cause in which so able, so zealous, and so eloquent an advocate has failed.

There is something in the very fact that I venture to address you upon a subject which has occupied the attention of your learned body now for several years, which has been regularly referred to a committee of able and distinguished men selected from your number, which has been maturely considered and elaborately reported on by them, and has at length, after full opportunity for discussion and comparison of views, been finally disposed of by the adoption and publication of their report, which convicts me, in appearance at least, of presumption, and which is not likely to win me your favor in advance. It is, therefore, due to myself to say, that in asking your indulgence while I recall your attention for a few moments to a subject which has lost its novelty for you, I am acting under instructions from a body whose authority

I am not at liberty to disregard. On the first day of May last a resolution was adopted by the Board of Trustees of Columbia College, in the following words :

" *Resolved,* That the President be requested to attend the next meeting of the Convocation of the University of the State of New York, and to explain to that body how far the views of the Faculty of Columbia College, in respect to the Metric System of Weights and Measures, are in accordance with those set forth in the report of a Committee made to the Convocation on that subject in August last."

This resolution imposes upon me a duty which I have no choice but to fulfil. I trust, therefore, that I shall not be suspected of intending any disrespect to the Convocation if, in discharging it, I give expression to views considerably at variance with those which have received the sanction of this body.

ACTION OF THE TRUSTEES AND FACULTY OF COLUMBIA COLLEGE ON THE SUBJECT OF METROLOGICAL REFORM.

The interest of the Trustees of Columbia College in the Metric System of Weights and Measures, dates back to a period preceding my own connection with that institution. In the minutes of the Faculty of the college, I find it stated, under date of April 8, 1864, that a resolution calling the attention of the Faculty to the subject had been passed by the Trustees on the Monday preceding ; and this resolution is recorded

under a date a little later, in the following words,
viz.—

"*Resolved*, That in view of the important interna-
tional movement in progress for the purpose of estab-
lishing a uniform system of Weights, Measures, and
Coins for the civilized world, it be referred to the
Faculty of the College to prepare and submit to this
Board a memorial to the Congress of the United
States, on behalf of the College, expressing its sense of
the importance of the measure in question."

In compliance with the request of this resolution, a
committee was appointed to draw up such a memorial,
consisting of Professors McVICKAR, ANTHON, DAVIES,
LIEBER, and ROOD. Upon the thirteenth of May next
following this appointment, the committee reported a
memorial which was adopted by the Board of the
College, and ordered to be signed by the President and
laid before the Trustees.

In this memorial the metric unit of length is compared with the British (and American) unit; and the
metric system of derivation, by which the measures of
surface, capacity, solidity, and weight, are deduced
from the unit of length, is compared with the confused
system or lack of system which unfortunately pervades
the metrology of the English speaking nations; and on
both these points the opinion of the Faculty of Columbia College is expressed, with an emphasis which
argues deep conviction, in favor of the former. The

memorial goes further, and advocates certain views in
regard to the unification of the coinage of the world,
in which I should not be able fully to concur, but to
which I design to give no present attention, since the
coinage question is one which will probably be independently settled ; though when it is settled it will
doubtless be settled by making the weights of all
coins metrical.*

In the minutes of the Trustees of the college, I find
it recorded that the document here spoken of was read
before that body on the 18th of May, 1864, and that
an order was afterwards taken directing that it should
be forwarded to the Speaker of the House of Representatives at Washington, with a request that he
should lay it before the national legislature. This
memorial is spread at length upon the minutes both of
the Faculty and of the Trustees. Along with it appears,
in the latter, an independent memorial expressing
similar sentiments, which was adopted by the Trustees

---

* The question of the unification of coins has been the subject of several
international conferences and of much diplomatic negotiation. A definite
proposition regarding it was addressed to foreign governments, in 1870, by
the Government of the United States, through its ministers resident abroad,
which proposition may be regarded as still pending. This consideration alone
would suggest the propriety of refraining from the discussion in this place of
the various projects for such unification, which have been heretofore proposed.
The question, however, is complicated by considerations of great gravity,
which affect but slightly the simpler one of the unification of systems of weight
and measure ; and which render it expedient, if not indispensable, that the
two questions should be kept separate. See APPENDIX A.

in their own behalf, and forwarded to Congress by their order at the same time.

With all this history I had nothing to do. The whole transaction had been completed before I became a member either of the Faculty or of the Board of Trustees of Columbia College. I have recounted the particulars in evidence that Columbia College had taken her position on this question long ago. The duty imposed upon me is to state to you some of the reasons which have led her to assume this attitude.

And here I wish to be understood as not intending to assert that, among the members of the two bodies whom I, in a certain sense, represent here, there are not individuals who dissent from the views to which the majority have committed themselves. Among the Trustees, indeed, I know of none such ; but your own records show that there is at least one distinguished dissentient among the Faculty. If there are others, they, as well as myself, are members of this convocation ; and they may, and perhaps will, speak for themselves. I am not aware that there are any, but my impression may be mistaken.

THE PREVALENCE OF A PARTICULAR SYSTEM LESS IMPORTANT THAN THE ADOPTION OF A COMMON SYSTEM.

To proceed now to the matter in hand, it is, in the first place, a fact particularly noteworthy, that the Trustees of Columbia College, in their resolution of

April, 1864, did not mention the metric system in so many words, nor propose to memorialize Congress in favor of any particular system of weights and measures designated by name. What they asked was that the college should express its sense of the importance of the creation of a *uniform* system of weights and measures for the use of the civilized world. If they have been led to believe, as I think they have, that such a uniform system can be reached in no other way than through the ultimate adoption of the metric system; and that whatever may be the differences of opinion as to this matter just here and now, such a result most inevitably will be reached in this way sooner or later; it is impossible that they can regard without concern the exertion of any influence which may serve sensibly to retard an event deemed by them so desirable, and which however retarded is, in their opinion, sure to come at last.

But still the question raised by them, and the question first in order before us now, is not, whether it is expedient that we should forthwith adopt the metric system; it is rather, whether it is worth while to try to secure a common system; because, if this is not so, there is nothing left to talk about. Nor yet, supposing that this first question is settled affirmatively, and we agree that a common system of weights and measures would be worth having if we could get it, can we make the question of the metric system even the second question in order; for the second question should be, what are the efforts, of those which we are able and willing to make to secure the desired object, in attempting which we may be encouraged by the hope of success ; and what are those which we may as well take warning in the outset to avoid, as certain inevitably to fail.

THE MAIN OBSTACLE TO METROLOGICAL REFORM CONSISTS IN THE INCOMMENSURABILITY OF UNIT BASES.

Now, in looking at the different metrological systems at present in use in the civilized world—and the number is very encouragingly smaller at present than it was even twenty years ago—we shall see that the great obstacle in the way of the practical unification of systems is not the mere fact of difference in the absolute magnitude of the standard units whether of weight or of length or of surface or of capacity ; nor the mere fact of difference in the names by which we distinguish these units and their multiples or subdivisions ; nor the mere fact of difference in the arithmetical law by which the several denominations of weight or measure are related to each other ; nor yet the greater or less degree of exactness with which the base of any system may conform to any dimension in nature variable or invariable ; though these are matters concerning which a great deal of breath is wasted in the discussions which go on about the metric system ; the difficulty which stands out so prominently as to dwarf all the rest to insignificance, is the fact that the standard units of these several systems are practically incommensurable. I say *practically* incommensurable—I am not using terms with strict scientific severity—but the difficulty is as serious as if the incommensurability were actual and absolute.

To illustrate what I mean: the Austrian foot measure, for instance, exceeds our own by 361 ten-thousandth parts ; or, expressed in inches, it is equal to 12 British or American inches and 4332-10000ths of an inch. We might reverse the mode of presentation, and say that the American foot is 0.96516 of an Austrian foot, or 11.5819 Austrian inches. The fractions here given do not express the exact relation. We may run the decimals on for half a mile without reaching the end ; but they go far enough for the purpose of my illustration.

To transform, therefore, a value expressed in one of these measures into an equivalent value expressed in the other, is an operation laborious and irksome. But the arithmetical disadvantage is by no means the whole, or even the greater part, of the evil which this state of things produces. A much more grave consideration is the fact that it interposes an effectual bar to the intelligent interchange of thought. It renders it impossible for an American to converse understandingly with an Austrian on any subject involving quantities of any description. It makes it impossible for an American to derive instruction from an Austrian book or magazine or journal where quantities are mentioned ; or an Austrian from an American. This is an enormous evil, and as it exists not in this quarter only, but everywhere, the world has crying need of its removal. It is the evil of which, first and chiefest of all, the advocates of metrological reform desire to be rid. And yet, unless I greatly misunderstand the purport of much of the reasoning I hear going on upon this subject, the very fact that this abominable evil exists, the very fact that there is something to be got rid of, something that we want to be rid of and something which we ought to be rid of, is made an argument why we should not try to be rid of it at all. If there are any to whom this argument is satisfactory, with them of course the case is closed, and upon such nothing that I can say will produce any impression. There are some, probably, not of this class, and they may be disposed to consider with me what means there are by which we may be relieved of this artificial obstruction to intelligent communication with other peoples.

METHODS POSSIBLE OF OVERCOMING THE OBSTACLE.

Three methods present themselves. Continuing the illustration furnished by the example just presented, they are these : We may adopt the Austrian unit ; Austria may adopt ours ; or both nations may adopt a third, incommensurable with either. Any one of these expedients will reconcile us with each other ; that one will be chosen, if any, which, besides doing this, will do most to promote the grander object of an universal accord among nations.

Now, if we go over to Austria in this matter, we lose more than we gain ; and in speaking of loss and gain here, I mean loss and gain as it respects our international relations, and not as it respects our internal or domestic affairs ; (I am setting aside altogether, therefore, that inconvenience to ourselves at home, which would exist temporarily in consequence of the incommensurability of the new unit with the old—an inconvenience on which your able reporter has so feelingly dwelt as likely to result from our giving up the foot for the metre)—we lose then I say, more than we gain by this concession, because we fall out of harmony with the great British Empire for the sake of securing harmony with a people who occupy, as Mr. Webster more forcibly than politely remarked, only a small patch of the earth's surface, compared with our vast domain.   On the other hand, if Austria yields the point, she will fall into harmony, it is true, with us and with Great Britain, nations however from whom she is geographically separated very widely ; but this advantage will be gained at the expense of discord with the several peoples who lie between her and the British Islands ; all of whom have embraced the metric system, or adopted metric values in their own.   If, however, thirdly, both of us adopt the metre, America will not only be in accord with Austria, but with the greater part of continental

Europe at the same time ; and as Great Britain has given decisive indications of a disposition to become metric also, the probability is that, soon, the whole civilized world, unless Russia and the Scandinavian peoples continue to be exceptions, will have but a single system of weights and measures.

SIMPLICITY OF THE PROBLEM OF UNIFICATION WHEN UNIT BASES ARE COMMENSURABLE.

Now if this difference between us and Austria had not been *such* a difference ; if the unit, for instance, of Austria, had been we will say a cubit, and such a cubit (for there have been many of them) as to be exactly equal to eighteen British inches ; the evil of incommensurability would not exist and the question of unification would simplify itself materially. We might both of us agree to adopt a measure of six inches as our common unit; and this we might call a span, abandoning both the cubit and the foot. There would be some, no doubt, who would lament over the loss of the "short and sharp Saxon word, foot," and would find no sufficient comfort even in "span," which though Saxon and short is not so sharp ; but I think that if a change of this simple description, affecting only names and modes of division and not actual values at all, would bring us into harmony with any great people, and still more if it would bring us into harmony with all the world, we should do it notwithstanding. This illustrates the

immense difference as it respects facility of solution
which the problem of metrological unification would
present, if the unit bases of the existing national
systems were commensurable with each other, as
compared with what it is now.

With the great empire of Russia, we have, indeed,
precisely such a point of contact in our two national
systems of length measure, as that which I have just
been imagining. It is matter of familiar history, that
in the year 1698, the singularly enterprising and ener-
getic emperor, PETER *the Great*, passed, incognito, in
the train of one of his own embassies, first into Holland, where he engaged himself as an operative ship-
carpenter, and labored for months with remarkable
assiduity; and secondly, in the following year, into
England, where he became finally master of that important branch of constructive art; and where subsequently, having made himself known to the then
reigning King, WILLIAM III., he received from that
monarch every honor due to his exalted station. Returning to his own country, he took with him a number of British ship-builders and other artificers, whom
he paid with liberality and employed in the navy yards
which he immediately proceeded to found. It was
probably a consequence of his own industrial education
in England, and of the British predilections of the
advisers and practical assistants whom he took home
with him to carry out his plans, that he was induced
to modify the length of the *sagene*, the standard length

unit of the empire, so as to make it commensurable with the British foot. Since early in the eighteenth century, therefore, the length of the legal standard unit of Russia has been seven feet, subdivided into three archines of twenty-eight inches each; the archine being the unit of common life, just as in England, the yard is the legal, and the foot the practical, standard. This state of things admitted of the introduction between ourselves and Russia of a common unit of four inches in length, which would have been just the seventh part of an archine and the third part of a foot. But within the last forty years a simpler solution has been found, the Russian government having introduced a subsidiary unit, one-seventh of a sagene in length, which is called by a name approaching as nearly to the English word foot as the vocal organs of the people will allow.

It is unfortunate that a reform so well begun was prosecuted no further. This is the only particular in which the Russian metrological system has anything in common with ours, or with any other existing. And, from a careful examination of all the systems of weights and measures established by law, and at present in use among civilized peoples, I have been unable, except in this single instance, to discover between any two of them, any feature of commensurability, whether as it respects weights or measures of length or surface or volume, which has not been introduced by legislation since this century began.

RECENT ENCOURAGING PROGRESS OF METROLOGICAL REFORM.

Such legislation has, however, within the period here indicated, done so much toward removing this chief of all obstacles to the attainment of a common system, as greatly to encourage the friends of metrological reform in the hope of an ultimate and complete attainment of the object in which they are so deeply interested. It is worth remarking, furthermore, that every such change has thus far consisted in replacing the values of the weights and measures in common use by other values adopted from the metric system. I will enumerate further on the most important of these changes; pausing here only to present a single example to show how, in some instances, the great evil of incommensurability has been got rid of without the adoption of the metric system in all its details.

In the republic of Switzerland, previously to the year 1851, there prevailed a considerable diversity of systems of weight and measure in the different cantons. The most important may be said to have been those of Berne, Zurich, Basel and Lucerne—I will confine myself to the first two named. In Berne, the *fuss* (unit of length) was 11.546 inches; in Zurich, it was 11.812 inches. In Berne, the *pfund* (unit of weight) was 18.642 ounces avoirdupois; in Zurich, it was 18.347 ounces. In Berne, the *maas* (measure of liquid capacity) was 1.766 quarts; in Zurich, it was 1.918

quarts. The measures of dry capacity were very various ; but those which are most easily comparable were the *maas*, in Berne, which was 0.3876 of a bushel, equal to 1.550 pecks ; and the *viertel*, in Zurich, which was 0.5826 of a bushel, or 2.330 pecks. Zurich had, in fact, four different viertels for different substances ; the measure given above was for wheat. By a law passed the 23d of March, 1851, it was decreed that, after the 31st of December, 1856, the legal unit of length throughout the republic should be the *pied* (foot), having the length of .exactly thirty centimetres ; and that (abandoning the old duodecimal subdivision) this should be decimally divided to tenths, hundredths, and thousandths. Multiples of the unit allowed were the *brache*, 2 feet ; the *aune*, 4 feet : the *toise*, 6 feet ; the *perche*, 10 feet ; and the *lieue*, 16,000. For the pfund was substituted the *livre* of 500 grammes ; and for subdivision it was left optional to use the binary system of half-pound, quarter-pound, and so on, or to employ the decimal. For dry capacity, the measure established was the *quarteron*, equal to 15 litres ; and for liquid capacity, the *pot*, equal to one and a half litres. For the last fifteen years, therefore, the weights and measures of Switzerland have been commensurable with those of the neighbors with whom she is in most frequent intercourse, and who have all of them, more or less completely, adopted the metric system.

DISCORDANT LAWS REGULATING DERIVATIVE DENOMINA-
TIONS INCREASE THE DIFFICULTY OF TRANSFORMING
VALUES.

But though the incommensurability of the unit
bases is the principal source of difficulty in effecting
transformations of value from one metrological system
to another, the law of derivation of the other denominations from these bases is not a matter of indifference
as it respects this operation. This, again, can be illustrated by an example which the commensurability of
the Russian standard unit of length with our own
enables us to derive from the itinerary measure in
common use in the Russian empire. This itinerary
unit is the *viersta*, anglicized verst, which is equal to
500 sagenes. Now there is no difficulty in converting
versts into British feet, since (the sagene containing
seven feet) 500 sagenes are 3,500 British feet, or the
half of 7,000 British feet ; so that we only have to
multiply the versts by seven, annex three zeros, and
take half the result. But if we wish to transform a
distance expressed in versts into miles—a mile being
the unit of our own itinerary measure—the process is
not so simple. There are 5,280 feet in a mile, and if,
in order to avoid this troublesome divisor, we attempt
a reduction through the intermediate denominations,
we encounter such relations as $5\frac{1}{2}$ yards, or $16\frac{1}{2}$ feet,
making a rod, which are more troublesome still. Now

if, instead of the present totally indefensible series of relations between our higher and lower denominations of length, we had something a little more sensible, or if not sensible, at least not worse than the Russian (for the Russian, with its ratio of seven to one, is only a little less bad than ours), the problem before us would admit of a solution comparatively simple. Suppose, for example, that a mile were made, as it might be, without harm to anybody, a round five thousand feet. It is true that "the old familiar mile of 1,760 paces," which your reporter seems to cherish so fondly, "would be gone ; and," as he correctly remarks, " the distance from Albany to New York—one hundred and forty-five miles—would be known to us" as something quite different, say something like one hundred and fifty-three sensible miles. In this case the verst would be 3,500-5000, or seven-tenths of a mile, and the transformation would be effected by the use of a very small number of figures. It would not, perhaps, be worth while to adopt the new value of the mile here suggested, for the sake of being able merely to convert miles to versts, or versts to miles—a thing we have to do too seldom to make it a matter of much interest to us ; but it *would* be quite worth while to make it for the sake of facilitating the thousand other calculations which we are continually called upon to make, involving reductions between the higher denominations of length and the lower.

ABSURDITY OF THE NUMERICAL RELATIONS BETWEEN THE
DIFFERENT DENOMINATIONS OF MEASURE AND WEIGHT
IN THE UNITED STATES.

Whether we ever adopt a new linear base for our
metrological system or not, every consideration both of
logic and of convenience demands that we should reform the absurd numerical relations in which our different denominations, especially of length, surface and
capacity, stand to each other. Mention has been made
of the mile, the rod, the yard, and the foot. Along
with these we may take also the chain of sixty-six
feet, divided into links of which each one is seven
inches and ninety-two-hundredths, while itself is the
eightieth part of a mile.

As to surface, our square yard is nine square feet ;
our square rod is $30\frac{1}{4}$ square yards, or $272\frac{1}{4}$ square
feet ; and the acre, our agrarian unit, is 160 square
rods, or 4,840 square yards, or 43,560 square feet. It
would be difficult for human ingenuity to contrive
anything more inconvenient or less rational than this.

As to capacity, whether liquid or dry, though the relations of the several denominations to each other are
tolerably simple, yet their relations to the standard of
length, which is, or ought to be, the fundamental base
of the system, are as abnormal as it is possible to make
them. Our gallon is 231 cubic inches = 0.13368 of a
cubic foot ; and our bushel is 2,150.42 cubic inches =
1.244456 cubic feet. And our unit of weight has no

relation whatever expressible in simple numbers which the mind can grasp, to our measures of capacity or of length ; for while it is customary to say that a cubic foot of water weighs one thousand ounces, the relation established forty years ago by our bureau of weights and measures, under authority of law, makes the gallon measure of distilled water at the temperature of maximum density, viz., 39°.8 F., and at thirty inches of the barometer, to weigh 58,372.1754 grains, or 8.3388822 commercial or avoirdupois pounds ; and requires also that the standard or Winchester bushel of 2,150.42 cubic inches shall hold, under the same circumstances, 543,391.89 grains, or 77.627413 avoirdupois pounds* ; from either or both of which determinations it appears that the weight of a cubic foot of water, at maximum density, is only 998.0667 ounces, instead of 1,000 ounces ; while, if we take the water at the ordinary temperature of the atmosphere, say 62° F., as prescribed by the British statute on the subject, the cubic foot weighs but 997.172 ounces.

Here are irregularities and imperfections, to the correction of which it would be well if we would address ourselves, in our own immediate interest and that of our people at home merely, and without reference to our relations with other peoples. Exact calculations, for instance, in which weights are to be deduced from volumes, or volumes from weights, are effected under our system only at the expense of much weary labor,

---

* The determination of the bushel, however, by weight, allows it only 2150.4 cubic inches. See APPENDIX B, *Note* 1.

of which the necessity is artificially laid upon us by this tyranny to which we are born. For rude calculations we call, indeed, the cubic foot of standard water one thousand ounces, or sixty-two and a half pounds ; but for any delicate determination, we must take the cubic foot at 436,654.1952 grains, and the cubic inch at 252.6934 grains, numbers which, at whatever inconvenience, the man of science finds himself continually obliged to employ in multiplication and division, to the great waste of his time and expense of his strength.*

IS IT POSSIBLE TO SECURE GENERAL COMMENSURABILITY OF UNIT BASES?

Now before directly considering the special question whether the metric system of weights and measures ought to be adopted in this country, it is proper to consider, and it seems to me to be. a duty to consider, whether it is possible for us to contribute anything to the important object of bringing the bases of the metrological systems of the world into relations of commensurability. If these bases can be made commensurable, we shall have accomplished something almost as important as to have established absolute identity ; and, indeed, under these circumstances, identity, it may easily be believed, will not be slow to follow. Who can doubt, for instance, that Switzerland, having adopted values for her units which are in extremely simple relations to the metric units, will sooner or later

* See APPENDIX B, for an examination of the probable accuracy of these values.

adopt the metric units themselves. It is what the states of Northern Germany recently resolved to do after an experience very similar. These several states had all, previously to the formation of the Zollverein, their independent systems of weights and measures. When that treaty was entered into, the importance of a common system for custom-house purposes was promptly perceived ; and hence a Zollverein pound was adopted, having the weight of five hundred grammes. Several of these states, among them Prussia, Baden, Hesse-Darmstadt and Wurtemberg, found it expedient at a later period to adopt this weight for their domestic as well as for their external commerce. And after the Austro-Prussian war of 1866, and the formation of a closer union between the states north of the Maine under the name of the North German Confederation, the desirability of a common metrological system for all the members of the confederation seemed so great, that a law was finally passed by the Reichstag, which was publicly proclaimed by the king in August, 1868, by which the metric system is adopted in full, and made the legal system for North Germany from and after January 1, 1872. As the states of southern Germany were no less advanced than those of northern, in the measures they had taken previously to their absorption into the empire, for the assimilation of their weights and measures to those of the metric system, there can be no doubt that the law of 1868, just mentioned, will be extended over them also.

Thus it appears that when different metrological systems approach each other so far as to become commensurable in their fundamental units, there is a drift towards identity which becomes at length irresistible.

THE WORLD WILL HAVE A COMMON SYSTEM OF WEIGHTS AND MEASURES ; THE CHOICE MUST LIE BETWEEN OUR OWN AND THE METRIC.

Are we willing to do anything to bring our own system into relations of commensurability with those of the rest of the world ? If so, the question second in order comes up, what efforts are there in our power to make, which are likely to advance the object, and what are likely to be fruitless? It may be well to state, in the very outset, so as to bring the really vital question directly before us, that except the metric system and that which we use ourselves, no other existing and no other likely to exist, can be advocated as having the least claim to become the system of the world. One of these, therefore, must sooner or later prevail ; for no man not totally regardless of the history of the past, and not absolutely blind to what is taking place under his own eyes in the present, can possibly pretend to believe that the world is to be forever without a uniform system of weights and measures. At the universal exposition of 1867, in Paris, thirteen measures of length from different countries were exhibited under the name of foot, or its equivalent ; but among these there were only eight values essentially different ; and two of these were metric. Yet after giving some attention to this subject without pretending to exhaust it, I have found more than one hundred foot-measures, each differing more or less from all the rest in value, which have been in use at one time or another at one part or another of Europe. Similar remarks might be made of the units of weight and capacity. There has therefore been large progress made toward uniformity, and the most important steps and the most significant steps are those which have been taken within our own century. We cannot suppose that this progress is going to be arrested at the point which it has now reached. Of the two systems therefore just now indicated as the systems between which the world must choose, unless in regard to this matter it shall henceforth stand still forever, one or the other must sooner or later prevail. Which shall it be ? Which is it likely to be ?

ABRUPTNESS OF THE INTRODUCTION OF THE METRIC SYSTEM INTO FRANCE.

At the close of the last century the metric system was thrust upon France, under circumstances of disadvantage and with an imperfect success which Mr. ADAMS has very eloquently described in his able report of 1821, which you have caused to be reprinted. Though the commission by which the system was matured was as far international as it was possible in the then existing political and military condition of

Europe to make it, representatives being present not only from France, but also from the Netherlands, Denmark, Sweden, Spain, Switzerland, Sardinia, Rome, and the Cisalpine and Ligurian Republics; yet no government except the French spontaneously adopted, and endeavored to apply in practice, the results of their labors. The conquests of the first empire carried the system forcibly into the Low Countries, into portions of Germany, into Italy and into the Iberian peninsula; but the difficulties which it met with there were in general greater than at home; not only because the manner of its introduction did violence to men's established habits of thought, but because its existence was a badge of subjugation and a perpetual reminder of the national humiliation of those who were compelled to use it. These all with one accord, therefore, took advantage of the downfall of the empire, to throw it promptly off. Nor even in France, presented as it was to the people without any adequate education as to its characteristic features, or any sufficient allowance of time to permit them to become familiar with its details, was it established without a struggle against inveterate habits and rooted prejudices continued through more than a quarter of a century.

SUBSIDENCE OF OPPOSITION TO THE SYSTEM.

Long before the termination of this struggle, however, the aversion to the system in the countries foreign to France to which it had been carried during the empire, began sensibly to subside, and in the Netherlands whose intimate relations with France had caused it there to take a deeper root than elsewhere, it was actually re-established as early as 1817. It was re-established, that is to say, in all particulars except the nomenclature ; but while the metric units and the metric decimal relations were adopted, the ancient names of weights and measures were retained. Belgium, however, which was for a time a part of France, employed the nomenclature also ; the old names were preserved in Holland till April, 1869, when they were at length abandoned.

The condition of things, therefore, at the time when Mr. ADAMS wrote may be thus described. A quarter of a century had passed and yet the system was not yet firmly established in its own home ; it had been rejected generally by the neighboring peoples who had tried it ; and its chances of success in the eyes of the disinterested spectators of the experiment appeared, as may be gathered from Mr. ADAMS's own report, to be as nearly as possible at zero. Since that period just one half a century has passed, and the aspect of things has bravely changed. One third part only of this period sufficed for the subsidence of all the imputed disaffection of France ; and in the Netherlands, as we have seen, this disaffection, if it was ever strong, died out much earlier. From the year 1837 onward, the people of those two countries have not only been reconciled to the system, but have been warmly attached to it. The neigh-

boring peoples upon whom it had been early imposed
by force, and who had indignantly thrown it off, have
all voluntarily re-adopted it. That early attempt to
coerce them into its acceptance, while it roused every
instinct of their natures to resistance, had at least the
effect to educate them to a knowledge of what it was.
And the acquaintance which they thus formed of its
merits, produced, when passion had subsided, its
natural result in the re-establishment of the system by
their own free choice.

ACCEPTANCE OF THE SYSTEM BY EUROPEAN STATES SUCCESSIVELY.

From information obtained at the universal exposition of 1867, where a special pavilion was set apart for
the display of the standards of weight, measure, and
money of all nations, officially authenticated, I am able
to state some particulars as to the progress of this
great movement toward metrological uniformity.
First in the order of time after France, Holland, and
Belgium, came the kingdom of Greece, which adopted
the metric system of weights and measures as early as
September, 1836. A little earlier than this (March,
1833) had been formed the customs union or Zollverein, which I have already mentioned, among the
German States, embracing, originally, Prussia, Saxony,
Bavaria, Wurtemberg, and Hesse Cassel ; but ulti-
mately including all the states of the German
Bund except Austria, Lichtenstein, Holstein, the two

Mecklenburgs, Hamburg, Lubeck, and Bremen. This was established for the purpose of doing away with the serious obstruction to commerce interposed by the existence of numerous and neighboring custom-house frontiers ; but in order that it might not create as much trouble as it removed, it adopted, as I have mentioned already, a common unit of weight, having the metrical value of five hundred grammes, or half a kilogramme. This, which first came into use on the first of January, 1840, was found so convenient that it subsequently became the national as well as the international pound in several of the principal states and many of the smaller ; as, for instance, in Prussia, Wurtemberg, Baden, and Hesse Darmstadt.

In some, as in Baden and Hesse Darmstadt, metrical values were also given to the unit of length, which was still called the *fuss*, though the old *fuss* was abolished ; and the duodecimal subdivision was at the same time abandoned. Others of these states which still retained the values of the fuss to which they were accustomed, perceiving the great superiority, in respect to convenience, of the decimal over the duodecimal ratio, also abolished the inch, and divided the foot into tenths and hundredths. Among these may be named Bavaria, Prussia, and Wurtemberg. In Italy, a law of 1845 brought the metric system fully into force in the kingdom of Sardinia from the first of January, 1850. It was established in the Pontifical States soon after ; and, by a law passed July 28, 1861, was extended over the entire Italian peninsula, and also over Sicily from and after the first of January, 1863.

Very shortly after the revolutionary excitements of 1848, the empire of Austria, by treaty with Prussia, became connected with the Zollverein, and introduced the metrical weights and measures of that union into all the custom-houses of her extended frontier ; nor has the course of events, military or political, of recent years, produced in that empire any change in this respect. The action of the North German Confedera- tion, by which the metric system in full became the system of all the Northern states, for domestic as well as for external uses, took place in 1868, as has been already mentioned. In the mean time, other states, not connected with the Zollverein, began to fall in with the drift now becoming so general. In 1851 took place the legislation in Switzerland above described, which gave to that confederation a metric system of weights and measures after 1856.

In 1852, Denmark adopted the metric pound of 500 grammes, decimally divided. In 1855, Sweden, with- out changing the values of her standard units, intro- duced partially the principle of decimal derivation for the inferior and superior denominations ; and this, by more recent enactments, reduced in 1865 to a single comprehensive law, she has extended through her whole system. In the following year, 1866, Norway, in regard to this matter, followed the example of her sister kingdom. In Spain and her colonies, the metric system was established by law in 1859, the names of the units having been partially transformed to bring them into harmony with the language—the metre being called the *metro*, the litre the *litro*, etc. In 1864, the metric system was established in Portugal ; and in the same year a law was passed in the principality of Roumania prescribing the use of the system in that principality from and after the first day of January, 1865. Even Turkey has recently made a beginning toward bringing her system into harmony with that which is now so rapidly becoming the system of all continental Europe, by giving to her unit of length, the archine, the value of seventy-five centimetres, or three-quarters of a metre. In Great Britain, by an act of Parliament passed in 1864, the metric system is legalized, though it has not been made compulsory. In 1868, another act, making the metric the exclusive system for Great Britain, passed its second reading in the House of Commons, having thus reached a stage of legislation where the final passage of a bill is commonly regarded as assured, when it was withdrawn by its originators, as yet premature ; but the fact of this remarkable success is a signal evidence of the state of opinion among the enlightened classes of the British people,* and a plain premonition of what Great Britain will sooner or later do.

---

* For a more full account of British legislation in regard to the Metric System, see APPENDIX C.

LEGISLATION IN REGARD TO THE SYSTEM ON THE AMERICAN CONTINENT.

In our own country the use of the metric system in business transactions was legalized by act of Congress of July 27, 1866. Another act, which was passed almost simultaneously with this, provided that postages should be charged in accordance with a scale of metric weights ; a letter weighing fifteen grammes or less to be chargeable with but one rate of postage. This provision of law, which was practically in favor of the people who use the post-office to a sensible degree, fifteen grammes exceeding the previously legal postage weight by nearly thirteen grains, or about one-seventeenth part, was, as I am informed, by the effect of a statute passed the following day, quite unintentionally repealed. This second act was designed to regulate postage with foreign countries, and it provided that, for postal purposes, one half ounce avoirdupois should be deemed and taken to be the equivalent of fifteen grammes. The department has applied this provision to the act of the preceding day ; so that we have Congress going through with the solemn farce of enacting that the limiting weight of a single letter shall be fifteen grammes, but then that these fifteen grammes shall be deemed and taken to be only half an ounce.

Undoubtedly the stationery on which these provisions of law were written must be deemed and taken to be a dead loss to the nation. But could anything more forcibly illustrate the liability to error and confusion arising out of diversity of systems of weight and measure, than this example wherein we see even our highest legislative body, when entangled in the maze, incapable of making laws to express its own intentions?

In South America, the metric system has been adopted by Brazil (to take full effect in 1873), in the Argentine Republic (1863), in Uruguay (1862), in Peru (1863), in Chili (1848), in Ecuador (1856), and in New Granada (1863). In North America, it was established by law in Mexico in 1856. According to the best authorities I have been able to find, the total population of Europe approaches 298,000,000, of whom about 135,000,000 have already accepted the metric system in all its details, or have given to all the standard units of their own systems, metric values. Add to these 25,000,000 more in Mexico and South America, and we have a total of 160,000,000 of civilized people in Christian lands who are irrevocably committed to the metric system ; while a considerable proportion of the rest have made progress toward this system by adopting metric values in part, like Denmark, and Austria, and Turkey; or by adopting the decimal law of derivation without as yet the metric values, like Sweden : while there are seventy millions more, the people of the British Islands and of the United States, who have made the use of the denominations of the system lawful in all business transactions within their territory.*

---

* For more exact statistics on this subject, see APPENDIX D·

PAST METROLOGICAL REFORMS NOT ATTRIBUTABLE TO POLITICAL CAUSES, BUT TO THE FORCE OF PUBLIC OPINION.

All this has been accomplished by the pressure of public opinion; it has been distinctively a movement of the people and not of governments; it is a social rather than a political phenomenon. When the metric system was first introduced into France, the pressure came from above, and was resisted by those upon whom it pressed. The people did not understand the system and they did not want it. In the discussions which we hear going on about us concerning it at the present time, the opponents of the system seem constantly to assume that the same plan is to be pursued to-day; and that there exists somewhere an insidious design to force the system upon peoples whether they like it or not. That, I take it, is not the spirit of the modern propaganda. Neither the British people nor the American people are expected to accept this system unless they think it best; but the presumption of some of us is that they will sooner or later think it best.

CAUSES OF THE INDIFFERENCE TO METROLOGICAL REFORM IN THE UNITED STATES.

But why, it may with justice be inquired, are our people so far behind those of the continent of Europe in appreciating the value of the metric system? This is to be accounted for by the same reasons which make them comparatively indifferent to the existence of *any* international system of weights and measures. In a large country like ours, widely separated from the rest of the world, the inconvenience of metrological diversity is immediately and personally felt by the individual citizen only on rare occasions ; when, for example, he travels in a foreign country, or when in his own he meets a foreigner raw to our institutions, or when he attempts to obtain some exact information from the publications of other countries; while the disadvantages to which it daily subjects him operate in a manner so indirect, and are mixed up, too, with so many other matters, that he fails to connect them with their causes. We can easily understand the state of things which would exist if we had no public standard of weights and measures at all ; and if every tradesman made his own system, and sold his customer, say, so much for so much. This plan is illustrated in Diedrich Knickerbocker's account of the dealings of the early settlers of the Nieuw Nederlandts with the Indians— "every Dutchman's hand weighed a pound, and every Dutchman's foot weighed two pounds." The inconveniences and uncertainty of trade would be only a little less if, instead of having as many systems as there are tradesmen, we should have as many as there are villages. If, for instance, while a man can get on very comfortably among his immediate neighbors, he finds himself, on driving four or five miles, entirely at sea on the subject of quantities, he will be scarcely able to prosecute any business of magnitude without an amount of trouble and confusion quite intolerable. Enlarge the communities within which common systems prevail, and separate them more widely from those which employ different systems, and the evils which, in the original supposition, embarrassed individuals, now affect the transactions which take place between these communities. Operations are larger, and they are mainly conducted by a particular class ; but the misapprehensions, the delays, and the increased expense attendant on these operations, are charged, like the customs duties, upon the whole community, without their being clearly conscious of the fact. Our custom-houses, and our great importing houses, are compelled, by the diversity of weights, measures, and moneys with which they have to deal, to employ an immense staff of computers, whose sole business is to effect transformations of values upon the invoices of the commodities which pass through their hands ; and the salaries of all these employees are undoubtedly paid by the consumers of the commodities.

Now, on the continent of Europe, where, in the central part at least, the territories of independent states have heretofore been small, while the population is dense, the evil of a multiplicity of systems of weight and measure, and of custom-house lines occurring every ten, twenty, or thirty miles, has been felt as, of course, we can never feel it ; and, therefore, there is nothing surprising in the fact that the people of those states have perceived the need of a common system to be pressing, when we were not thinking of the matter at all. Nor is it any more surprising that, in looking about for a common system, and finding the metric system to be an existing system, and a good system, and, above all, an available system, and the only one apparently available for the purpose, they should have seized upon it, and legalized it, and made it permanent, without too anxiously concerning themselves with the questions whether the metre would not have been better if it had been a little longer or a little shorter, or if it had represented something different from what it does represent, or whether, in fact, it does, after all, really represent any thing at all.

UNIVERSAL PREVALENCE OF THE SYSTEM IN EUROPE INEVITABLE.

Considering, therefore, the nature of the causes which have induced one hundred and thirty-five millions of the people of Europe to adopt the metric system ; and considering furthermore that in Denmark, Sweden, Norway, Austria, and Turkey, we have fifty-five millions more who have shown, by their legislation, their appreciation of the merits of this system, or of the principles on which it is founded ; it may, I think, be safely said that the universal extension of this system over the continent of Europe is only a question of time.

THE SCIENCE OF THE WORLD FAVORS THE METRIC SYSTEM.

Besides the causes which I have mentioned, out of which the important changes I have just described have grown, it is to be remembered that there are other influences of a very powerful description actively at work to recommend the metric system to the favorable consideration of the peoples which have not yet received it. The principal of these peoples are the English speaking nations, and the inhabitants of the Russian empire. Now, for a very long period, it has been true that the great body of the scientific men of our own country, of Great Britain, and of Russia, have been thoroughly impressed with the value of the metric system; and many of them have been constantly in the habit of using it. That there has been here and there a dissenter may be admitted. Here, as elsewhere, *exceptio probat regulam.* But a dissenter who, like Sir JOHN HERSCHEL, holds that the system is good, but that the base ought to have been a ten-millionth part of something else, rather than of a quadrant of the meridian, is not much of a dissenter after all ; and one who, like Prof. PIAZZI SMYTH, bases his metrological theories on religious grounds, and prefers the pyramid-inch as his standard, as a matter of conscience, is not likely to concentrate around him a very powerful party of opposition.

Scientific associations in the countries just named have memorialized their governments in favor of the metric system. The British Association for the Advancement of Science has done this repeatedly, and the Imperial Academy of Sciences of St. Petersburg has done it likewise. In the year 1866, the National Academy of Sciences of the United States, on the report of a committee having Prof. HENRY, of the Smithsonian Institution, at its head, (a committee which had had for two years the subject in its charge), addressed a memorial to the Congress of the United States, expressing the sense of the Academy as to the importance of establishing an international system of weights and measures, and recommending the metric system as the best existing.

The scientific journals, throughout the world, give evidence of the growing practice of scientific investigators of using metric values in their experiments, in their calculations, and in their writings. This began in Germany very early. I find it to be true of POG-GENDORFF's *Annalen* so long ago as the year 1800. At the present time it is next to impossible to find any other system of weights and measures but the metric so much as occasionally named in any of the publications devoted to physics and chemistry in all Germany. Almost the same thing is true of the scientific periodicals of Russia, of Austria, of Denmark, and of Sweden. I have very recently looked through the principal journals of this class, published in the countries just named, and what I assert of them, I assert from personal knowledge. Men of science have adopted this system, not only because of their approval of its principles, but because it is a labor-saving machine of immense capabilities. If you look into our own scientific journals and those of Great Britain, you will find that, what has just been remarked of the journals of the continent, is true to a considerable extent of them also ; and to an extent constantly increasing. Our analytic chemists use the metric system altogether; and with our physicists its use is becoming every day more general. With the science of the world on its side, therefore, the metric system has a powerful ally, which, added to the influence of the material interests enlisted in its favor, must make its final triumph inevitable.

THE SOCIAL SCIENCE ORGANIZATIONS FAVOR THE SYSTEM.

It will be understood that the scientific associations and the scientific men to whom reference is here made, are those who deal with the exact sciences, or employ themselves with material nature. Truth is the object of their search: with the uses of the truth discovered, or its relations to the human race, they do not concern themselves. There is, however, another class of inquirers, one to which I have earlier referred, whose influence on the question before us is destined to be powerfully felt, who have created in these modern times a new science of their own, taking as their subject precisely what the former class omits—the relations of truth to humanity. In the scope of their inquiries, they are most widely comprehensive, embracing equally all truth—the moral and the psychological no less than the physical. They call their science social science; it might be called the philosophy of philanthropy, for its object is to discover and remove the causes of human wretchedness, whether they be material, political, mental, or moral ; and to place the human race in circumstances where it may work out for itself a destiny the noblest of which it is by nature capable. It is not through a merely native taste or bias, that these men pursue the science they have created. Their science is to them more than a love— it is a religion. Their impelling principle is deeper than enthusiasm—it is an earnest sense of duty. These men, therefore, belong to that class whose characters command the highest respect, and whose opinions carry the largest weight among their fellow-men. They, too, like others, have availed themselves of the powerful machinery of associated effort. In many enlightened lands, social science associations hold their periodical meetings, and by means of the reports of their discussions scattered among the people through the public prints, and the wide circulation of their own memoirs and journals, powerfully impress the public mind. Within the past ten or fifteen years, there has sprung up, in addition to the national organizations here referred to, an international social science association also, composed of men of every land, many of

them men whose names have a world-wide celebrity.
The last meeting of this influential body was held in
1867; the next is to assemble in England, during the
ensuing fall. Among other measures adopted at the
meeting preceding the last, was the appointment of a
committee to draw up a complete code of international
law, to be presented for acceptance to the governments
of all nations, and to be binding upon all such as shall
assent to its provisions. The code is to comprehend
two grand divisions, presenting the rights and duties
of nations, first in peace, and secondly in war. The
first of these divisions, relating to peace, is now complete, and will be presented to the association at the
ensuing meeting. I am authoritatively informed that
it embraces provisions making the metric system of
weights and measures the common system for all the
nations accepting the code; and there can be no doubt
that the association will cordially concur with their
committee as to these provisions. The social science
associations may therefore be regarded as another
powerful influence, silently acting throughout every
corner of every civilized land, and throwing its whole
strength in favor of the universal adoption of the
metric system of weights and measures.

THE SYSTEM HAS BEEN RECOMMENDED BY INTERNATIONAL CONFERENCES.

There are other influences co-operating with this, in
which the political principle is combined with the

social. The importance of endeavoring, in some way or other, to arrive at a common system of moneys, weights, and measures, has been felt by governments to be sufficient to justify the calling of international conferences to discuss this very thing. Now, though these conferences have not resulted as yet in bringing actually to pass the object for which they were summoned, still they have furnished an independent and an important indication of the extent to which public opinion everywhere is turning toward the metric system, as destined inevitably to be at length the system of all mankind. For while the money question has invariably elicited a large variety of opinion, and while an agreement of all the delegates upon any one proposition for the unification of the coinage of the world has been found extremely difficult, if not impossible, to secure, yet, as to the question of weights and measures, there has been no difficulty whatever. At the Paris conference of 1867, for example, twenty-two nations were represented, including the non-metric nations, Russia, Austria, Sweden, Norway, Denmark, Great Britain, the United States, and Turkey. The report of the committee in favor of the metric system was an admirable document, drawn up by the celebrated DE JACOBI, of St. Petersburg; and it received the absolutely unanimous concurrence of all the delegates of all the nations. It is impossible to regard a phenomenon of this kind without seeing in it both an indication and an influence— an indication showing the march of opinion hitherto, and an influence which cannot fail to be felt in accelerating this march.

THE INTERNATIONAL STATISTICAL CONGRESSES HAVE ADOPTED THE SYSTEM.

But there is still another and a still more powerful influence, uniting, like that last mentioned, the social and political features, which has been gradually taking shape and gathering strength within the past twenty years, which is also destined to act powerfully in favor of the speedy creation of an international system of weights and measures, and which is already committed in advance to the metric system for that purpose. In explanation of this remark I will state that, about twenty years ago, there was assembled at Brussels, on the invitation of the government of Belgium, a convention which assumed the name of "The First International Statistical Congress." This body consisted of two hundred and thirty-six members, who were about equally divided between Belgium and foreign countries, thirty-five being delegates appointed by governments. This first convention, held in 1853, has been followed by six others ; of which the second was assembled in Paris in 1855 ; the third at Vienna, in 1857 ; the fourth at London, in 1859 ; the fifth at Berlin, in 1863 ; the sixth at Florence, in 1867 ; and the seventh at the Hague, in 1869. The spirit in which these great international assemblages originated, is explained in the following brief extract from the report of Mr. S. B. Ruggles, of New York, the delegate from the United States* to the convention of 1869, at the Hague, recently published by order of the United States Senate. "The distinguished promoters of the first congress, at Brussels," says Mr. Ruggles, "had seen enough of modern statemanship to know that the government of nations, in their present state of material progress, cannot be wisely conducted without a thorough knowledge of quantities;" and that the systematic collection and philosophical arrangement of the "quantities" needed for showing the general condition of nations, is "an indispensable preliminary to any recommendation by an international congress of any measures seeking to promote the general welfare." In accordance with this spirit, "the official report (or ' *compte rendu*') of the congress at Brussels shows its labors to have been largely devoted to the scientific analysis of quantities, in subjects interesting to all nations, to be used as a basis of a uniform system of inquiries, in actually collecting the necessary facts." And in like manner all the succeeding congresses have devoted themselves sedulously to the labor of bringing together every description of facts obtainable, in regard to the actual wealth, the productions natural and artificial, the condition of industry and commerce, the character of the social institu-

---

* Mr. Ruggles also ably represented the United States in the fifth congress, at Berlin.

tions, and other matters of kindred interest, relating to
the various peoples who make up the population of the
globe.   The results of such inquiries could only be
made available for any useful purpose, on the condition that all the " quantities " so ascertained should
be reduced to a form in which they could be compared ; on the condition, therefore, that they should
be expressed in denominations of the same system of
weights and measures ; and, accordingly, it has been
urgently recommended by all these congresses that all
statistical statements everywhere should be made in
terms of the metric system.   The seventh and most
recent of these assemblies, moreover, inaugurated a
work which, if efficiently prosecuted, will be in honorable harmony with the magnificence of the idea which
originated these congresses of the nations.   The nature
of this work is thus stated by Mr. RUGGLES : " On the
last day of the session, Dr. ENGEL, the distinguished
director of the statistical bureau of Prussia, presented
to the body, in general assembly, a plan of great com-
prehensiveness and importance, which had been matured after full discussion in the appropriate section,
and conversation with most of the governmental delegates.   It provides for a full and systematic explora-
tion of the whole field of international statistical inquiry, which is divided for that purpose under twenty-
four different heads, each to be the subject of a separate investigation by the delegates or members from
some one of the nations to be selected, and which is to

embrace the statistics under that head of all the nations. This great work, if fully carried out, will furnish, in convenient encyclopedistic form, a systematic series of carefully prepared reports on most of the subjects of highest interest to the statesmen and legislators of the different nations. Editions of at least two thousand copies of each report are to be published in uniform octavo volumes, under regulations presented in the plan, which was unanimously adopted by the congress, with strong expressions of approbation."

Without the metric system, the vast mass of information thus collected would be unavailable—the encyclopedia would be illegible. This system has, therefore, thus become something more than a mere instrumentality in the service of statistical science ; it has become even an integral part of the science itself. Henceforth the two are so irrevocably wedded that they can be separated no more forever.

IMPORTANCE OF THE INTERNATIONAL STATISTICAL CONGRESS.

The " International Statistical Congress " may now be regarded as an established institution. Its eighth meeting in the order of succession will be held some time during the course of the year 1871, and probably in St. Petersburg. Already the influence of its deliberations, of the published results of its labors, and of the spirit of comprehensive statesmanship which it has inculcated and fostered, is beginning to be sensibly felt, and with each successive decade of years it will be felt with a power continually increasing, in educating the minds of the peoples, and in moulding the counsels of governments into harmony with the great principle that nations only then consult their truest interests when they consult the common interests of humanity.

EARLIEST SUGGESTION OF THIS INSTITUTION BY MR. ADAMS.

The germ idea of an agency which, with time, has developed itself into a power capable of controlling, and destined so largely to control, the future of human history, is to be found in the report of Mr. ADAMS to the House of Representatives of the United States Congress, made in 1821, which has been already cited in this paper. Though this report discouraged the adoption of the metric system by Congress, and though its reasonings had the effect undoubtedly to impress the popular mind in this country with the conviction that the introduction of the system into these states is hopeless, yet the author himself was as deeply imbued with admiration of this system, considered as a scientific creation, as the warmest of its advocates ; and no one felt more profoundly than he, how great would be the boon to humanity, if one uniform system of weights, measures, and moneys could be made to prevail everywhere throughout the world. In the view of his large and statesman-like intellect, very many of the embarrassments which attend intercourse between nations, spring from the selfish and narrow legislation which looks only to the immediate interests or convenience of particular communities, and disregards the results to the great family of man. To him, all nations and all races are brothers by blood, inheriting the earth as their common patrimony ; and though, in the existing state of human society, it is necessary that the artificial lines which divide states from each other should be preserved, it is eminently desirable that, for as many purposes as possible, they should be kept out of sight. He therefore proposed that the President of the United States should be authorized to invite the governments of the several states having diplomatic relations with that of the Union, to appoint delegates to a congress of nations, charged with the duty of deliberating upon measures likely to be promotive of the general welfare ; but, foremost and especially, upon the possibility of establishing a uniform system of weights and measures for all mankind. That this important proposition was productive of no immediate result, is attributed by Mr. RUGGLES, and with apparent justice, to the political condition of Europe during all the earlier portion of this century ; and especially to that compact of political rulers for the suppression of liberal thought, and the stifling of all freedom of political discussion, which the momentous events of recent history have since shattered, known as "The Holy Alliance." Happily, however, at length, to use the vigorous words of Mr.

RUGGLES, "We find the germ of the general convention, planted by the far-seeing sagacity of Mr. ADAMS, in 1821, though slumbering for a generation beneath the surface, actually fructifying in 1853, when the first general assemblage of nations by government delegates, and really international in its objects, was convened in Brussels."

From this epoch dates a new era in the history of the world's legislation. For the enlarged views of the reciprocal duties, as well as of the true interests of nations, in which this great general movement originated, are destined, through its instrumentality, to impress themselves more and more completely upon human institutions; until statutes shall at length cease to be monuments of ignorance, prejudice, or ignoble jealousies, and the aim of all laws shall be the greatest good of the greatest number. One most important result has already been secured by the action of these congresses; in that, so far as the science of statistics is concerned, so far, we may even say, as the successful conduct of governmental administration is concerned, it has made the metric system of weights and measures a system of universal necessity, and rendered a familiar acquaintance with it absolutely indispensable to every statesman, every publicist, every teacher or student of political economy, and every enlightened lawgiver throughout the world.

ALL CAUSES CONSPIRE TO RENDER THE ULTIMATE
TRIUMPH OF THE SYSTEM INEVITABLE.

It thus appears that there are powerful, permanent, and all-pervading influences steadily at work to advance the cause of metrological reform ; and that these influences conspire to forward the movement in the direction which it had already spontaneously taken— that is to say, toward the ultimate prevalence of the metric system of weights and measures over every other. It further appears that the actual progress which the movement has made since the century began, has by far exceeded anything which could have been reasonably anticipated, and has been sufficient to justify the most sanguine hopes for the future. When we consider, for example, that, at the close of the last century, the simple measure of length called the foot had not less than sixty different values still—probably many more—actually in use in different parts of Europe ; and that, in 1867, at an Exposition in which the measures of all the world were all brought together, there could be found only eight of this discordant class still surviving ; argument would seem to be needless in behalf of a cause which is so manifestly making its own way unaided. Whether our own people are to be participators in this grand movement, which has already gone so far, is not with me a question of probabilities, but only a question of time. I expect very little to-day, and not much to-morrow ;

but beyond to-day and to-morrow there are other days coming, from which I expect everything. I know the strength of early associations and the power of rooted habits ; I know how fondly men will hug the evil which is familiar and reject the good that is strange. I know that the Greenlander greatly prefers his icy mountains to the coral strands of India. My inference is, that we must look to a generation which shall not be so mentally one-sided as ours ; a generation in whose training the good shall not be placed at so tremendous a disadvantage as it has been in our own ; a generation which shall bring to this great metrological question a judgment at once fair, candid, unbiased, and unwarped by the prejudices which mislead and bewilder us ; to pronounce the impartial decision for which, it must be sadly admitted, we seem disqualified ourselves. And such a generation, gentlemen, permit me to predict, will yet be born upon the American continent, if it is not born already.

·II.—*Objections to the Metric System considered.*

PRELIMINARY REMARKS.

In all that I have hitherto said, I have not dwelt for one moment upon the intrinsic merits of the metric system itself. I have not thought that necessary. I am addressing intelligent men who know the system, and who know that, for the whole circle of our dealings with quantities, it stands, for easiness of apprehension, for convenience of use, and for the degree to which it facilitates reductions, precisely where our Federal currency stands, among systems of money. The simplicity of the relations, moreover, by which it connects the measures of surface, of capacity, and of weight, with the linear base, is such as is nowhere else found ; and such as to make of the system a powerful intellectual machine, and an educational instrumentality of inappreciable value. All this I pass by. But I cannot pass so lightly by the objections which have been urged against the system, and of which, in my view, the importance has been, in most instances, exaggerated beyond all reason ; since, through the wide circulation of the report of your committee on this subject, the high authority of this learned convocation has been made liable to be popularly regarded as attesting their gravity. Consistently with the duty imposed upon me on this occasion, therefore, I cannot pass them by ; although the extent to which I have already trespassed upon your indulgence forbids that I should examine them with all the fulness that I could desire.

FIRST OBJECTION—THE UNIT BASE TOO LARGE.

We are told, then, first, that the linear unit of the system is too large. Too large for what? Too large, in the words of your committee, "to be apprehended by a young and uninstructed mind." This is something which I confess that *I* do not apprehend. A metre, I suppose, can be brought into the school-room; and can be seen without difficulty, even by a very small boy, from end to end. I remember, when I was a very small boy myself, seeing something brought in which was about as long as a metre; and if I did not apprehend it at the time, I was at least very apprehensive of it.

But Mr. ADAMS says the metre is too long for a pocket rule. "Perhaps," he remarks, "for half the occasions which arise in the life of every individual for the use of a linear measure, the instrument, to suit his purposes, must be portable, and fit to be carried in his pocket. Neither the metre, the half-metre, nor the decimetre is suited to that purpose." What then would Mr. ADAMS have? Would the foot rule fit into a man's pocket more conveniently than the decimetre? Does any man carry a foot rule in his pocket in any other than a folding form? And cannot a folded metre be carried in the pocket as easily as a folded foot? I at least find it so ; as this rule proves, which I here present you. But since we have not yet adopted the metre as our unit, and since, after all, in spite of what Mr. ADAMS says, or what anybody else says, it happens to be notorious that a foot is *not* the measure which, " for half the occasions which arise in the life of every individual," is the most useful ; the portable measure which we commonly find in men's pockets is a tape measure of a yard or a fathom in length, put up more compactly than is possible for any rule, whether long or short.

As to what *ought* to be the value of the standard length-unit, opinions differ. The British standard is a yard. The Russian is the sagene, more than twice as long. Capt. PIAZZI SMYTH almost fanatically attaches himself to the inch, a measure which he believes with implicit faith to have been divinely given to CHEOPS, builder of the great pyramid, and again to MOSES in the wilderness ; and in what he, no doubt, regards as the great work of his life, he uses no other to express the largest dimensions.

But it is also said that there are things to be measured in the common affairs of life that are less than a metre. I should suppose so. There are likewise many things to be measured, less than a foot, or an inch. They measure these things in England even though the yard is their legal standard. In mechanical engineering, in France, the centimetre is the unit ; in physics the millimetre. It does not unfit them for these uses, that their names happen at the same time to be expressive of relation to the standard. The metric unit of weight in commerce is the kilogramme ; in analytic chemistry and pharmacy it is the gramme. The metric unit for dry measure is the hectolitre ; for liquid measure, it is the litre; the metric agrarian unit is the hectare ; the metric itinerary unit is the kilometre. It is, in fact, one of the merits of the system, that while, like all other systems, it allows any denomination to be made a unit measure for special purposes, yet it allows also instantaneous transformations from one denomination to another without changing a figure, but by the simple removal of a point. This cannot be done in non-decimal systems. The inch, for example, is with us the unit of the mechanical engineer and the draftsman. The rod is the farmer's unit of distance. But to reduce inches to feet you must divide by twelve, changing all your figures ; and to reduce rods to feet you must multiply by sixteen and a half. This plan does not seem to me preferable to the metric. When your committee say that in their opinion " other units besides the base-unit should be used, as secondary bases for collections of numbers," I agree with them. It is what, in the employment of the metric system, I have always been in the habit of doing myself. But if, when they say this, they mean to say that values expressed in units of these secondary bases ought *not* to be transformable by the simplest processes possible into units of the standard base, my impression is that they will fail to carry the world along with them.

SECOND OBJECTION—THE DECIMAL DIVISION TOO DIFFICULT.

Another serious difficulty is started, of an educational character. Ten, it seems, is a difficult number to grasp ; and one-tenth part is a still more difficult fraction. We can never know anything about one-tenth, " until we have divided the unit into two equal parts, into three, into four, and so on up to ten." Since this is the case, it is melancholy to reflect how much more objectionable is our actual system of weights and measures than the metric ; since it will be necessary to divide the foot into two equal parts, into three, into four, and so on all the way up even to twelve, before the faintest conception of an inch can begin to dawn upon our minds ; and when we turn our attention to the pound and the ounce avoirdupois, the formidably protracted extent of this unavoidable operation becomes quite disheartening. Still, however grave this business of ten may be, I suppose that our children must some time or other know something about decimal arithmetic; and they will have to know something about it whether they learn the metric system or not. If they know it, they know the system, all but its nomenclature; if they don't know it, then I can conceive no educational machinery better suited to make them know it, than the visible magnitude of the

metric measures placed before their eyes. The question is not whether we shall teach the metric system to
babes, but whether we shall teach it along with the
arithmetic, and as a part of the arithmetic, which boys
must learn at any rate. The objector does not apparently discover that his argument is no less damnatory
to our Federal currency than to the metric system ;
yet my observation in the streets of New York satisfies
me that *gamins* of very tender years, without having
enjoyed the advantages of scholastic culture, or having
been carefully and systematically carried through the
mental operation of dividing the unit into two parts,
into three parts, into four parts, and so on up to ten,
acquire an acute appreciation of the relative value of a
dime stamp and a nickel.

THIRD OBJECTION—THE DECIMAL DIVISION UNSUITABLE
FOR PRACTICAL PURPOSES.

It is objected again that, while the decimal ratio is
infinitely more favorable to calculation than any other,
yet for sensible objects, and for the daily purposes of
life, the binary subdivision is to be preferred. If so,
then let us use the binary so far as convenience may
demand. There is no need on this account to reject
the decimal, which for purposes of calculation is of
priceless value. No harm is going to arise from employing both. We divide the dollar certainly into
halves and quarters, to our great convenience ; and
the decimal system of the Federal currency is none

the worse for that. We used to divide it into eighths and sixteenths even ; and Mr. ADAMS says that, if the Spanish mint had not furnished us with coins representative of these values, we should have been obliged to coin them ourselves. Yet within ten years after Mr. ADAMS wrote, we had effectually swept out all this fry of foreign coinage, and nobody now perceives the want of it. Again, the Swiss pound is half a kilogramme. Take half any number of Swiss pounds and you have kilogrammes. Double the number of kilogrammes and you have Swiss pounds. The Swiss, moreover, use both the decimal and the binary subdivision. I presume they would not do this if they did not find it for some purposes useful, as we do in our Federal currency. The Swedes, some fifteen years ago, introduced the decimal subdivision, but they still retain some binary relations. Some such binary relations are recognized also by the French law ; but it does not therefore follow, as your committee infer, that this fact "must give rise to much confusion." Neither is it true, as they also maintain, that we cannot adopt the essentials of this system without "adopting it as a whole and excluding every other :" by which I understand them to mean that we shall not even adopt metric values for our units, as Denmark has done, and as Austria has done, and as Turkey has done to a certain extent, without adopting the nomenclature throughout, and sternly prohibiting the use of all binary division ; or that we shall not adopt if we please the decimal relations, as Sweden has done, without adopting either metric values or the nomenclature; nor adopt the metric values and the decimal system complete, and yet reject the nomenclature, as Holland continued to do for half a century, and has only ceased to do within the last two years. Surely, things that other people have done, we may do ; nor is there going to be, as your committee apprehend, any "fierce conflict" about the matter, nor any need to talk about "the spirit of a free people," or to insist on the fact that Americans are not habituated to "blind obedience to imperial edicts."

THE SPIRIT OF METROLOGICAL REFORMERS MISAPPREHENDED.

Can it not be understood that nobody of the great party who are seeking metrological reform and perfect international accord on this important subject, is bigotedly devoted to the metric system for its own sake ; or resolutely determined to yield nothing that is in it, or to accept nothing that is not in it, on any consideration whatever ? Their battle is for a *common* system, be that what it may ; but if they believe that that common system will be found at last to embrace the main features of the metric system, they are not to be told that they shall have nothing else, if anything else superadded to it will make it either theoretically or practically better. Mr. ADAMS wrote fifty years ago. What he wrote seems to have impressed your

committee much more forcibly than all that has happened since. But the world has moved since the time
of Mr. ADAMS ; and it is perhaps not quite in order to
tell us that, if we think it a good thing to divide by
ten, we shall never be permitted to divide by any other
number so long as we live. The first great point to
be secured is commensurability of unit bases. That
point once gained, the battle is substantially over. As
for nomenclature and subdivisions, however important
these matters may be, their importance is secondary,
and they may be attended to afterwards.

THE DANGER OF JUMPING AT CONCLUSIONS.

Sweeping propositions are rarely wholly true. It is
not a fact that binary subdivisions of weight and
measure are always necessarily the best. In small
dealings, the convenience of buyers and sellers is best
consulted, when the multiples and submultiples of
quantities correspond with the multiples and submultiples of coins. If a pound of any commodity costs
twenty-five cents, it would suit all parties who use the
Federal currency better, if we could divide the pound
evenly into five parts, than it does now to divide it
into four. Nothing is more certain than that quantities bought and sold, and the instrument of purchase
and sale, should be subject to the same law.

COMMENSURABILITY OF UNIT BASIS FIRST TO BE SOUGHT.

But, as just remarked, the first point and the great point to be secured is, commensurability of unit bases. This can be accomplished, if we please, with great facility. Our foot differs from three decimetres by a very inconsiderable fraction—less than two-tenths of an inch. If we make this slight change in the length of our foot, we are in harmony with nearly all of continental Europe. As for the other measures, they present no difficulty when the measure of length is once adjusted ; for measures of length determine the dimensions of permanent constructions, while pounds and gallons are for ascertaining quantities of substances usually perishable. Men are disposed, therefore, to adhere with more obstinacy to their measures of length than to those either of weight or of volume.

PAST INSTABILITY OF BRITISH MEASURES AND WEIGHTS.

Mr. ADAMS's report shows that, in the past history of England, nothing has been more unstable than the value of the pound, the bushel, and the gallon. There was a time when the gallon of liquid capacity contained only 216 cubic inches—in one sense a judiciously chosen value, since it was just one-eighth part of a cubic foot. The dry measure gallon contained, at the same time, 264.34 cubic inches, corresponding to a bushel of 2,114.68 cubic inches. And there was a ratio connecting the liquid and dry measures, which was that of the specific gravities of wheat and Gascon wine. Mr. ADAMS is quite enamored of this duplicity, which extended to the weights as well, between which the ratio has been pretty closely preserved down to our time. But this liquid gallon went on, as Mr. ADAMS explains, to be successively 217.6 cubic inches, 219.43 cubic inches, 224 cubic inches, and finally, as with us now, 231 cubic inches. As to the bushel, it seems to have had all sorts of value. By statute of 1496, passed in the reign of HENRY VII., it seems to have been ordered that this measure should contain 1792 cubic inches ; but this statute was never carried out. There are two exchequer standards of this reign, one of 2124 cubic inches, and one of 2146 cubic inches, which latter is called the Winchester bushel. But then, under HENRY VIII., we have the large bushel of 2256 cubic inches, from which came the ale gallon of 282 cubic inches, so long in use with us. A bushel afterwards appeared, of 2148.5 cubic inches ; and subsequently the Winchester bushel was found to have somehow worked its way up to 2150.42 cubic inches : at which value it was, in 1701, made the standard in England, and so became the standard with us, as it continues to be yet.

SUMMARY PROCEEDING OF THE BRITISH PARLIAMENT ON
THE SUBJECT.

But just three years and a half after Mr. ADAMS so
strongly expressed his regrets at the destruction of the
beautiful " uniformity of proportion " contemplated
by the theory of the British measures, the British
Parliament took this whole business in hand.   Instead
of improving the capital opportunity afforded them of
correcting the irregularities which his report signalizes,
they quietly struck out of existence every measure of
capacity in use, whether wet or dry; and established
the system of *imperial* measures, wherein the bushel
contains 2218.1907 cubic inches, and the gallon 277.-
2738 cubic inches, to be used equally for commodities
of all descriptions.   This was a tolerably formidable
change and a tolerably sudden change; but it occa-
sioned no insurrection; nor did the people even run
after the carriages of the ministers, shouting " give us
us back our bushel;" as we are told they shouted in
1752, " give us back our eleven days," when the Gre-
gorian calendar was first introduced into England.
Changes of metrological systems, then, are possible,
and are possible even for us, without provoking
" fierce conflicts."   All that is necessary is that the
people should know what they are, and should feel
that they are desirable.   .

THE COMMITTEE OF THE CONVOCATION THEMSELVES ADVO-
CATE REFORM.

Your committee themselves are not adverse to all change. There is one modification of our system of weights which they actually propose to our acceptance. The recommendation is made moreover so impressively, out of a sense of " duty, plain and imperative," that for one I was prepared for something startling ; for at the very least a proposition to do away forever with the perfectly unnecessary Troy pound. I am compelled to confess my disappointment. The proposed innovation so solemnly introduced is explained in the following words: "In analyzing these weights, it is found that the ounce in the apothecaries' weight and the ounce in the weight Troy are identical, and that each exceeds the ounce avoirdupois by its eighty-three-thousandth part very nearly; hence, if the ounce Troy, or the apothecaries' ounce, be diminished by its eighty-three-thousandth part, the result will be the ounce avoirdupois, or the one-thousandth part of the weight of a cubic foot of distilled water, and then these three weights will have a common unit."

I have pondered this passage profoundly, but I have not been able to see my way to the bottom of it. It has been my lot to be compelled to transform Troy ounces into avoirdupois ounces very frequently; but I have always found the difference to be 42.5 grains, while it is here apparently hardly six one-thousandths of one grain. Presuming, however, that something may have been intended which is not said, and that, at any rate, it is designed somehow to make the Troy and apothecaries' ounce equal to the avoirdupois ounce, I accept this proposition as a concession so far as it goes to the cause of uniformity and simplicity ; but I ask what justification can exist, after abolishing the smaller denominations, which alone are used by the jewellers and dealers in bullion, or even by the druggists (for the wholesale drug trade is carried on in avoirdupois pounds)—what justification can exist after this, for retaining the useless pound of twelve ounces.

I would point out further that, since the ounce, after being reduced by nearly its eleventh part, is still, according to the proposition of your committee, to consist of four hundred and eighty grains, the grain must accordingly be reduced in the same proportion; so that all the confusion which could arise in pharmacy and the trade in precious metals from changing the grain for the milligramme, whereby something might be gained, will be here introduced without gaining anything at all.

FOURTH OBJECTION——THE DECIMAL DIVISION HAS FAILED, AS APPLIED TO THE CIRCLE.

It has been furthermore urged as a fact very injurious to the pretensions of the metric system, that this system has never been permanently applied to the division of the circle, to which, if to anything, it ought to be peculiarly adapted. Those who use this argument ought to remember that the Arabic numerals, the symbols of algebra, and the division of the circle, are three things (and the only three things, I believe), which were the same for all civilized mankind when the metric system was created. To change the law of circular division was to introduce diversity where uniformity prevailed before ; and also to destroy the usefulness of a vast scientific literature which had been founded on the sexagesimal division. Yet the French did make the experiment of dividing the quadrant centesimally, both in tables and in instruments ; and what was thought of its convenience by the ablest astronomers and geodesists of that day may be inferred from the following incidental remarks of DELAMBRE, in his description of the operation of measuring the great French meridian arc. " Three of our four circles," he observes, " were divided into decimal grades or degrees, each having the value of $360° \div 400 = 0°.9 = 540' = 3240''$. This division is much the most convenient for the uses of the repeating circle, and would be equally so for the verniers of all instruments whatever. Many persons hold to the old system by habit, and because they have made no use of the new ; but no one of those who have practised both, will willingly return to the old."

When the metric system shall be universal, it is probable that the decimal division will be once more

applied to the circle. Nothing could be less convenient than the sexagesimal which is now employed. And
in point of fact, this law of subdivision has been already abandoned for all values below seconds ; such
values being now invariably expressed decimally,
though, two or three hundred years ago, it was carried
to thirds, fourths, and even fifths, as may be seen in
any old astronomical work, or in DELAMBRE'S *History
of the Astronomy of the Middle Ages.* I regard this
objection therefore as without foundation.

FIFTH OBJECTION—THE UNIT OF LENGTH SHOULD BE SOME DIMENSION OF THE HUMAN PERSON.

But it is apparently a very strong point with most
objectors to the metric system, that our present measures of length have their representatives—the assumption is that they have their prototypes—in the dimensions of some parts of the human body. Thus, your
committee say, the foot " was undoubtedly adopted as
a standard of measure from the part of the body from
which it takes its name." Some foot was undoubtedly
so adopted, but what foot ? The Greeks used the foot
earliest, and the Olympic foot is said to have been the
measure of the foot of HERCULES. But there were foot
measures in use among them of several other magni-
tudes ; and while it is difficult to know with certainty
what any of them were, compared with ours, it is not
difficult at all to ascertain that they differed widely
among themselves. Thus the authorities state that the

Macedonian foot was 14.08 inches, the Olympian 12.14 inches, the Pythian 9.72 inches, and the Sicilian 8.75 inches. Here, in the earliest history of this measure, we have the largest room for choice. In more recent times, the diversity has been greater still. Thus in Italy the foot was, not long ago, 11.62 inches in Rome; 13.68 inches in Lombardy ; 23.22 inches in Lucca. In France, it was 9.76 inches in Avignon ; 9.79 inches in Aix-en-Provence ; 10.57 inches in Rouen ; 14.05 inches in Bordeaux ; while the *Pied du roi*, for France generally, was 12.79 inches. In Switzerland, it was 10.52 inches in Neufchatel ; 11.33 inches in Rostock ; 11.99 inches in Basel ; and 19.21 inches in Geneva. In the Spanish Peninsula, it was 10.12 inches in Aragon, and 10.96 inches in Castile. In Germany, it was 9.25 inches in Wesel ; 10.89 inches in Bavaria ; 10.998 inches in Heidelberg ; 11.45 inches in Göttingen ; and 13.12 inches in Carlsruhe. In the Netherlands, it was 10.86 inches in Brussels, and 11.28 inches in Liège. These examples will suffice, but there are plenty more behind.

It can hardly be supposed that all these measures were taken from the human foot ; it is hardly probable that any of those used in the later centuries were so. The name has been perpetuated from a very early time ; but the thing named has either lost by degrees its original value, or it has been arbitrarily changed. As to the origin of the British foot, it is pretty easily explained. There is no reason to doubt the account commonly given of the adjustment of the yard from the arm of HENRY I., in 1101. The foot is certainly derived from the yard, which has always been the standard of length in England, and is simply the third part of that measure. I know that we are continually told that our American foot is in length but a fraction in excess of the average foot of man. It astonishes me that any one who has two feet to walk on himself should ever entertain this opinion. The length of the human foot is given in the *Encyclopedia Britannica*, (authority Dr. THOMAS YOUNG) as 9.768 inches. Upon how large an extent of observation this determination is founded, is not known; but the question in issue is pretty well settled in the volume of " *Investigations in the Military and Anthropological Statistics of American Soldiers*, by Dr. B. A. GOULD," published, in 1869, among the Memoirs of the U. S. Sanitary Commission. Nearly 16,000 individual men, volunteers for the army, of very various races and nationalities, were subjected to measurement, of whom about 11,000 were white, and the rest colored. Dr. GOULD says : " The mean length [of the foot] was found for no nationality to exceed 10.24 inches ; and for none to fall below 9.89 inches ; the value for the total being 10.058 inches," or about a twentieth of an inch above ten inches. This approaches much more nearly to a quarter of a metre than to a third of a yard.

Let it be understood that nobody is objecting to the foot measure. It is a very convenient measure to have.

If it were slightly modified so as to be equal to three decimetres, it would be more desirable still; but it is quite unnecessary to defend it on the ground that it is the measure of the human foot; and it is judicious not to do so, because that happens not to be the case.

However, the facility of measuring off the yard on the arm is a fact which furnishes to the objector firmer ground. We *can* do that. Sir JOHN HERSCHEL's rule is: "Hold the end of a string or ribbon between the finger and thumb of one hand at the full length of the arm extended horizontally sideways, and mark the point that can be brought to touch the centre of the lips, facing full in front." Very well; now if you will carry the string or ribbon entirely across the lips, and mark the point that can be brought to touch the angle of the jaw or the lobe of the ear, you will have a metre. Or, if you carry the ribbon across the breast instead of the lips, and bring it to the point characteristic of that part of the person, you will have a metre once more.

The breadth of the palm is a decimetre; the breadth of the little finger at its extremity is a centimetre. A pace is an artificial step, and not a natural one; but suppose that it were natural for us to stride three feet, or suppose, at any rate, that we have learned to do so; and suppose that a metre is too large a step to be easily acquired; a pace is practically nine-tenths of a metre, and any number of paces are reduced to metres by dropping a tenth part. Thus, fifty paces are forty-five metres, and one hundred paces are ninety metres. This reduction is the simplest of all possible processes. Thus, I do not see that, by adopting metrical measures, we are going to be in the slightest degree disabled from finding, in the dimensions of our own persons or of our steps, all the means of effecting rough measurements which we possess at present ; and this objection falls to the ground.

SIXTH OBJECTION—AN OBJECTION OF THE COMMITTEE—WE CANNOT CONVENIENTLY DEAL IN ONE SYSTEM AND THINK IN ANOTHER.

But there is still another practical objection which is so perfectly well founded, that I hardly know what to say about it ; so that I am not sure that the truest wisdom in me would not be to let it alone altogether. It is the undeniable truth, that, *if we give up our present measures we shall cease to have them any longer.* " What follows ?" say your committee with anxiety ; " we have blotted from the mind of the nation the foot, and a knowledge of every measure into which it enters as a unit." This is evidently a serious business. It reminds us of the sad case of the lad, who, having eaten his cake, desired to have it again. The committee go on to explain that, instead of twenty-five feet we shall have to say something else ; and instead of one hundred and forty-five miles we shall have to say something else still. And exploring the extent of the calamity, the committee become gloomily figurative ; and, speaking with deep emotion of "the cubic foot, known wherever the English language is spoken," they tell us that this cherished object "is also gone, and in the twilight of its existence, we grope about for a substitute." I do not deny that this is eloquence ; but I respectfully submit that it is not argument. There cannot but be some of us who will consider that this tenderly lamented cubic foot, with its inconvenient numerical relation to the cubic inch of 1728 to 1 ; and its more inconvenient relation to the common unit of liquid capacity of 1728 to 231 ; and its even still more inconvenient relation to the unit of dry capacity of 1728 to 2,150.42, is very well out of the way.

I will not attempt to follow the committee further in their lament. But I cannot omit to notice, in passing, the perplexing embarrassment of the honest man who, setting out to purchase the convenient quantity of fourteen pounds of beef for his dinner, after there have ceased to be any pounds, is astounded at finding that he will be compelled to pay for the amazing number of grammes expressed by the figures six thousand three hundred and fifty-six ; or in case that he is bankrupted by this huge demand, will be permitted to compromise the matter only on condition of buying six kilogrammes, three hectogrammes, five decagrammes, and six grammes. I wish to present a parallel to this. I go to my tailor for a coat, and he states to me the price, in a sum expressed by the four digits named above, in the same order, viz., six, three, five, six.

The committee has given the general rule for reading concrete decimal numbers, as follows : " All the readings are made in the lowest unit." Hence, the cent being the lowest money unit involved in the price named, my tailor is under the necessity of informing me that I can have the coat for six thousand three hundred and fifty-six cents ; and it will not be lawful for him to vary the form of expression in any manner unless to say, by way of alternative, that he will give me the coat for six eagles, three dollars, five dimes and six cents.

I would, however, advise the unfortunate man who finds so much trouble with his marketing, not to buy his meat by the pound after pounds have gone out of date ; but to content himself with a round six kilogrammes, or, in case he is very hungry, say six and a half.

SEVENTH OBJECTION—THE ADOPTION OF THE SYSTEM WILL INVALIDATE LAND TITLES.

As it respects the objection that the introduction of the new measures would invalidate the titles to lands held under old surveys, nothing can be more imaginary. No legislation on this subject can be retroactive —it would not be constitutional if it were. The registry of deeds in the past would continue to have the same validity as now. In making a new deed in the future, nothing would be easier than to translate the language descriptive of linear and superficial dimensions from one form of expression to the other. Changes would thus come on gradually, as property should change hands. Deeds have to be made anew when sales are effected, and only then. The labor of making them in one form or the other is precisely the same.

NINTH AND TENTH OBJECTIONS—THE BASE NOT WELL CHOSEN, NOR CORRECTLY DETERMINED.

One final objection, or pair of objections allied to each other and closely connected together, I have reserved to be considered last. Some gentlemen honorably eminent in science have criticised the metric system on the ground that its base is not well chosen. This base purports to be the ten-millionth part of a quadrant of the terrestrial spheroid. But it is said the earth is not a spheroid, being rather an ellipsoid of three unequal axes; whence it follows that the meridians are unequal, and that the metre, if truly the ten-millionth part of one quadrant, is not a ten-millionth of any other differently situated in the ellipsoidal surface. The polar axis of the earth, on the other hand, is the common minor axis of all meridians; it is a magnitude entirely unique ; and, even if the earth were a true spheroid, there would be a higher degree of scientific fitness, there would be something on which the mind would dwell with more entire satisfaction, if we should take a fraction of that axis as the base of a system of metrology, rather than a fraction of any quadrant, or any other known magnitude. This is the view of Sir JOHN HERSCHEL and of Capt. PIAZZI SMYTH, and if the whole thing were to be done over again, it would probably be the unanimous view of the scientific world. But the matter has gone too far now to change the base. In the meantime, therefore, there is no impropriety in saying, that it is by no means yet proved that the earth is an ellipsoid. Neither, indeed, is it proved that it is a spheroid, if by that word is to be understood a figure geometrically true. What *has* been proved may be understood from the following succinct statement.

SKETCH OF GEODETIC OPERATIONS CONDUCTED HITHERTO.

There have been measured upon the surface of the earth, in all, excluding re-measurements, some sixteen meridian arcs. Most of these are very short, not exceeding three or four degrees in length, and generally less than two. The longest of them all is the Russian arc, of twenty-five and one-third degrees; and the shortest, the first Swedish arc, measured in 1737, by MAUPERTUIS, of fifty-seven and a half minutes. Two short arcs have been measured on the American Continent, one in Peru and one in Pennsylvania. The latter, only about one and a half degrees in length, was measured by Messrs. MASON AND DIXON, in 1767, without triangulation, and is esteemed of comparatively little value. The Peruvian arc, which is rather more than three degrees long, was admirably triangulated by BOUGUET and LA CONDAMINE, in 1735, and the two or three years succeeding. A short arc of about a degree and a half was measured at the Cape of Good Hope by LACAILLE in 1751. In this measurement, the effect of local attractions on the plumb-line was such as to lead to very erroneous conclusions. This arc has been recently re-examined and extended to more than four and a half degrees, by Messrs. HENDERSON and MACLEAR ; this operation bringing to light the causes which had vitiated the former. A long arc of twenty-one and a third degrees has been measured in India. With the exception of the Indian, the African, the Peruvian, and the Pennsylvanian arcs (the last hardly meriting to be included in the enumeration), all the rest are in Europe, and are embraced within limits of longitude differing, at widest, but about twenty-seven degrees.

DETERMINATION OF THE EARTH'S FIGURE FROM GEODETIC MEASUREMENTS.

Now supposing the earth to be a spheroid, it matters not, for the determination of its figure, what are the longitudes in which meridian measurements are made, provided the latitudes are different ; for on this supposition degrees in the same latitude are equal everywhere. Also, if the spheroid is oblate, the curvature in the higher latitudes will be less and the degrees longer than in the lower. Now, as in an ellipse, the linear amplitudes of any two arcs differently

distant from the apsides, along with the angles made
by the normals at their extremities, suffice to determine the axis and the eccentricity, it was to be expected that a comparison of any two properly selected
meridian arcs measured upon the earth's surface in
different latitudes, would furnish constantly the same
value of the polar and equatorial diameters, and the
same value for the compression of the poles.   But this
expectation has been singularly disappointed.   The
international scientific commission which, in 1799, fixed
definitely the length of the metre, in comparing the
French arc with the Peruvian arc, made the compression of the earth one 334th ; but Messrs. LAPLACE and
LEGENDRE, both eminent geometers, members of that
commission, by comparing one portion of the French
arc with another, made it, the first, 1-150th and the
second, 1-148th. DELAMBRE, one of the geodesists who
effected the measurement, deduced from his comparisons with the Peruvian arc, the value, one 312th, and
afterwards one 309th.   The French arc was subsequently extended southward nearly three degrees
more ; making a total length of twelve and one-third
degrees, when a recomparison with the Peruvian arc
by DELAMBRE gave a compression of one 178th.   The
effect of these differences of result upon the calculated
length of the quadrant of the meridian passing through
Paris would not be very great, upon the hypothesis
that the earth is really a spheroid ; for it happens that
the French arc is so situated as to give very nearly the

value of the mean degree, independently of the eccentricity. But if the earth is an ellipsoid, it is evident that it is entirely wrong in principle to compare two arcs with each other, when they differ materially in longitude.

INVESTIGATIONS OF GEN. T. F. DE SCHUBERT.

Now it is a part of the history of this subject that, in the year 1859, Gen. T. F. DE SCHUBERT, an officer of the Russian army of distinguished ability, after a laborious series of comparisons of several arcs combined two by two in all possible ways (the arcs were eight in number, and the combinations twenty-eight), found such remarkable discordances, that he felt himself forced to the conclusion that the earth is not spheroidal, but must be ellipsoidal in form. The compressions found by him varied, for instance, between the wide extremes of one 14501st, and one 116th; and the difference between the largest and smallest value for the polar axis amounted to 362,126 feet, or 68.584 miles.

Now observe what these deductions prove, and what they do not prove. They prove certainly that the earth is not a perfectly regular spheroid, and in this they are corroborated by other evidences; but they do not prove it to be an ellipsoid. The corroborating evidences just alluded to may be slightly glanced at in passing. In the first place, the successive degrees of the French arc do not increase, in going northward, in

the manner they ought if the meridian is truly ellip-
tical. And, secondly, it is true, that after that arc had
been extended southward, as above mentioned, to the
Island of Formentera, in the Mediterranean, the degrees at the southern extremity were found actually to
diminish in going northward, instead of increasing, as
in a regular ellipse they should have done.

Colonel EVEREST, also, the accomplished geodesist,
who executed the measurement of the northern section
of the great Indian arc, found that, when he compared
the northern half of the northern section with the
southern half of the same section, he obtained an
eccentricity of one 192d ; but that when he compared
the southern half of the northern section with the
whole southern section, the resulting ellipticity was
one 390th, or only one-half as great. The values of
the polar axis of the earth also, obtained from these
comparisons, differed by 67,106 feet, or about 12.71
miles.

These facts (and many like them might be stated)
are to be borne in mind in judging how far the method
of General DE SCHUBERT, with the data thus far gathered to go upon, is to be trusted. This gentleman,
concluding very properly that comparisons of arcs
measured in different longitudes are unworthy of confidence, resolved to deduce values of the polar and
equatorial diameters of each meridian, by such comparisons as that of Colonel EVEREST just described.
But here his material is at once largely reduced ; for

of the eight arcs employed in his previous comparisons, only three are long enough to permit the application of this method, viz.: the Russian, twenty-five and one-third degrees; the Indian, twenty-one and one-third degrees, and the French, twelve and one-third degrees. The British arc is now long enough to allow a fourth comparison (ten and a quarter degrees), but it is so nearly in the meridian of the French arc that it may better be treated as a prolongation of that. Gen. DE SCHUBERT divided each of his three arcs into two parts, as nearly equal as convenience would allow. From each he thus deduced a value for the major and minor axis of the meridional ellipse. If his hypothesis was true, the minor axes should have come out equal and the major axes unequal. The latter anticipation was realized, but the former only imperfectly so. The polar axis found from the Russian arc, compares pretty well with that found from the Indian; differing only about fifteen hundred feet, or rather more than a quarter of a mile; but the difference between the values of the same axis, as deduced from the Russian and the French, is fifteen thousand one hundred and sixteen feet, or nearly three miles. On account of this discrepancy Gen. DE SCHUBERT discards the French arc in this computation, and determines a value for the polar axis on the basis of the Russian and the Indian alone; giving, at the same time, quite arbitrarily, twice the weight to the former as to the latter. And with the axis thus determined and the aid of the Peruvian arc, he finds a third equatorial radius; which, combined with the Indian and Russian equatorial radii, enable him to place the axes of his imaginary equatorial ellipse. Finally, with the axes of the equator and their longitude, and also the equatorial eccentricity, he is able to compute the length of the equatorial radius corresponding to the French arc; and from that, the length of the theoretic French quadrant. Then, comparing this theoretic quadrant with the length of the same as deduced from the actual measurement of its ninth part, he feels himself justified in pronouncing the metre to be too short by the two hundredth part of an inch. I think it does not require a profound mathematician to see that the data on which this conclusion rests are too meagre to justify so important a deduction. The case is one to which Prof. HUXLEY's witty remark upon the power of the mathematics may be properly applied. "The mathematics," observes the Professor, "may be compared to a mill of exquisite workmanship, which grinds you stuff to any degree of fineness; but, nevertheless, what you get out depends on what you put in." And here it appears to me that we are not yet prepared to put in material enough to furnish us with a grist worth carrying away.

INVESTIGATIONS OF CAPT. A. R. CLARKE.

Prof. AIRY perceived the weakness of this method and pointed it out. He suggested an improvement on it which is worth more, and his suggestion was taken up by Captain A. R. CLARKE, an accomplished officer connected with the ordnance survey of Great Britain. This method consists in bringing together the latitudes, determined both geodetically and astronomically, of as many stations as possible, upon selected meridian arcs ; and then, all the elements of the problem being left indefinite, proceeding to ascertain what values given to the indeterminates will make the sum of the squares of the errors of latitude a minimum. He first presented his results to the Royal Astronomical Society, in 1860; and afterwards, having slightly modified some of his data, republished them in an appendix to a large volume issued in 1866 by the Royal Ordnance Survey. His last conclusion puts the metre in error one 172d of an inch. The number of latitudes employed by Captain CLARKE in this investigation is forty. Some slight variations made upon a portion of those in the Russian and the French arcs, between the first and the second determinations, amounting generally only to very small fractions of seconds, produced a sensible difference in the length of the polar axis, in the value of the compression, and in the computed error of the metre ; reducing this last from one 163d of an inch, which was his original determination, to one 172d, as

given above.   But Captain CLARKE himself regards
the data as entirely insufficient to make a correct
determination of the earth's figure a possibility.   His
own words are : " It would scarcely, I conceive, be
correct to say that we had proved the earth not to be
a solid of revolution.   To prove this would require
data which we are not in possession of at present,
which must include several arcs of longitude.   In the
mean time it is interesting to ascertain what ellipsoid
does actually best represent the existing measurements."   And having found this, he proceeds next to
apply the same method, *i. e.*, the method of least
squares, to the object of ascertaining, secondly, what
spheroid will best represent existing measurements ;
and he is brought thus to the conclusion that such a
spheroid is nearly as probable as an ellipsoid ; the
numbers representing these probabilities being 154
and 138 respectively (where the smaller number indi-
cates the greater probability).   We may admit then
that the ellipsoidal theory is slightly the more probable ; and with this preliminary we are prepared to
consider the two objections spoken of above.

CONSIDERATION OF THE NINTH OBJECTION RESUMED.

The first is, that the earth's meridians being unequal,
the ten-millionth part of a quadrant, even if we had
such a measure correctly, could be only the ten-millionth of one particular quadrant, so that the ideal of
a natural standard everywhere present and belonging

equally to all the world must be abandoned. Still it
cannot be denied that the quadrant chosen, though a
particular quadrant, possesses the essential property of
a standard, that is to say, invariability, quite as completely as if all the quadrants were equal. If this
natural standard were intended to be, or were capable
of being made, a standard of convenient reference, and
not merely a standard of value ; if, in other words, a
tradesman, suspecting his metre to be in error, could
adjust it by simply stepping out of his door and apply-
ing it to the earth's meridian ; there might be some
reason for complaint on the part of those, and they
would be the majority of mankind, whose distance
from the standard would deprive them of this facility.
This not being the case, no practical disadvantage
arises out of the inequality of the meridians, and it is
only the simplicity of the original conception that
suffers.

THE TENTH OBJECTION—DISCORDANCE BETWEEN THE ACTUAL AND THE THEORETICAL METRE.

The second objection to the base of the system—an
objection which is often urged in a tone which implies
that the objector regards it as nothing less than fatal—
is that the metre is not, after all, exactly the ten-millionth part of the particular meridian from which it
was derived. It is possible that it is not : nay, we
may safely assert that it would be nothing short of a
miracle if it were. We have glanced at the condition

of the problem of the earth's figure and magnitude in the hands of the geodesists. We have seen that every meridian measurement which has yet been made has served but to accumulate evidence that this figure is not geometrically regular, and is not probably, if words are to be applied with severe exactness, either spheroidal or ellipsoidal. It will easily be understood that a local irregularity actually affecting but a limited extent of a terrestrial arc, may, when it is allowed to give character to a whole circumference, lead to extraordinary conclusions; and we have seen the fact that it will do so, illustrated in the examples cited from Col. EVEREST and others. What hope can there be that the effects of such irregularities can be eliminated by an investigation which, however admirable in principle and however ably wrought out, rests on a comparison of only forty latitudes? Not a single geodetic measurement has yet been made in all the immense expanse of northern and eastern Asia, of northern and central Africa, or of Australia and the Australasian archipelago. Nor, except in the small Peruvian arc, and the still smaller Pennsylvanian, which latter does not count, has the great American continent made any contribution to the solution of the difficult problem under consideration. When we consider, therefore, that the introduction of minute corrections, amounting only to small fractions of seconds, into only a part of the data employed in Captain CLARKE's equations, suffices to modify the resulting dimensions of the earth to such an extent as to produce, as we have seen, a very sensible change in the calculated value of the error of the metre, I think that the assertion just now made will be admitted to be perfectly well founded, *i. e.*, that if the length given originally to the metre had been exactly the ten-millionth part of the Paris meridian, this result would have been neither more nor less than a miracle. I may further add that, even if the metre had been quite correct, its authors could not have known it to be so, and we should not know it to be so now. When measurements shall have been made in those vast regions just mentioned, which have not yet been attacked by the geodesists, and when, instead of forty latitudes, four thousand shall have been thrown into the hopper of Prof. HUXLEY's mill (though I confess that in such a case I should not be envious of the miller's task), we shall get out an inevitably different and a very certainly more satisfactory grist than has yet been ground for us. In the meantime, we may as well take the metre as we find it, and not concern ourselves about this Protean and microscopic fraction of error, which has so long been thrown up to it as a reproach.

EVERY ASSUMED " NATURAL STANDARD " LIABLE TO THE SAME OBJECTION, WHICH IS THEREFORE NO OBJECTION AT ALL.

It is a little remarkable that the objectors who find the error of the metre to be so grave a blot upon its

character, should nevertheless agree in urging us to
accept a standard derived from another natural dimension of the earth, equally invariable no doubt with the
quadrant, but at the same time equally unmeasurable
—the polar axis or the polar radius. This is a dimension of which the authorities give us as many different
values as they give of the quadrant; and of which
they are sure to give us a new one every time an addition is made, no matter how trivial, to the data from
which it is deduced. The values fluctuate perhaps
between narrower limits of variation; and if the ten-
millionth part of the earth's polar axis or the earth's
polar radius were our theoretic metre, the absolute
error of our practical metre would be probably rather
less in proportion to its length, than that of the metre
now in use. But the error would be there none the
less; for, as before, it would be nothing short of a
miracle if it were not; and between two errors, both
of them microscopic, and neither of them affecting any
conceivable human interest, I see for my own part
little to choose. If the advocates of the radius metre
could come to the defenders of the quadrant metre,
and say to them, "Here, you see our metre has no
error at all, and yours has one," the case would be a
strong one; but that does not seem to be the case.
Since these things are so, why then, you may inquire,
should we endeavor to fix our standard of length with
reference to either axis or quadrant or any other
dimension which we do not know, and which it is per-

fectly certain that we shall never be able exactly to
ascertain? That, gentlemen, is a question which you
may very well ask, but which I shall not attempt to
answer. I accept the metre as it is, not because it is
the ten-millionth part of the French quadrant (though,
according to Captain CLARKE, it *is* the ten-millionth
part of the quadrant passing through New York, within less than the ten-thousandth part of an inch), but
because it is the actual base of an admirable system of
weights and measures already in use among one hundred and sixty millions of people, rapidly growing in
favor among those who have not yet adopted it, and
destined in my belief to be sooner or later the system
of all the world.

NO COERCIVE MEASURES DESIRED IN BEHALF OF THE METRIC SYSTEM.

But, gentlemen, I do not expect that this system
will make its way in the world against the will of the
people of the world. I do not expect that our people,
and I do not desire that any people, shall be coerced
into receiving it by the force of " imperial edicts " or
by the terror of bayonets. What I do expect is, that
they will sooner or later welcome it as one of the
greatest of social blessings. What I do expect is, that
they will one day become conscious of the many inconveniences to which they are subjected from the
anomalous numerical relations which connect, or rather

we might say, disjoin, the several parts of their present absurd system ; inconveniences which they have learned to endure without reflecting on their causes or suspecting that they were unnecessary in the nature of things ; and that when fully at length awake to the slavery in which they live, they will burst its shackles, and rejoice in the deliverance which the metric system brings. This cannot take place, of course, until the people are thoroughly informed. There are influences, therefore, which are now only beginning to operate, which must first have their full course before the results I anticipate will make themselves manifest.

EDUCATIONAL INFLUENCES TO BE INVOKED.

The first and most important of these is the education of the young to a thorough understanding of this system, and a perfect familiarity with its practical applications. The metric system must be taught in all our schools. It ought of course to be taught there, as being the system actually in use among nearly or quite half the inhabitants of the civilized world already, and without any regard to the question whether it is to be ours or not. But it ought to be taught, too, with special reference to this question, in order that another generation may meet it and settle it intelligently. And I think I hazard nothing in saying, that when one generation shall have grown up, into whose minds this knowledge shall have entered along with the first rudiments of their learning, the question will no longer have two sides.

CUSTOMS DUTIES TO BE LEVIED ON THE BASIS OF THE METRIC MEASURES AND WEIGHTS.

But, in the second place, the system should be practically illustrated before the eyes of our people, by being introduced into our custom-houses, and made the guide according to' which duties are assessed and collected. This measure will disturb the habits of no one in the affairs of ordinary life. Importing and exporting merchants will interpose no objection to the change. On the contrary, they will welcome it as greatly diminishing the amount of computation which they are now compelled to make. It is, in fact, the complaint of Capt. PIAZZI SMYTH, that it was the pressure of the commercial class which came so near to making the metric system the exclusive system in England in 1868. Our tariff laws will require transformation ; but that transformation may be made without in any manner disturbing their essential provisions : so that no trouble need arise from this cause. What it is here proposed to do is nothing more nor less than what was actually done, some thirty years ago, by all the members of the German Zoll-verein. And though the state of things produced by it there will be superseded on the first of January, next, by the extension of the metric system in full over all the

component states of the late North German confederation, if not over the entire German empire ; yet it
will still exist in Austria, and will continue to exist in
that empire until she, too, shall adopt the same system
for her domestic affairs likewise.

PUBLIC SURVEYS TO BE CONDUCTED IN ACCORDANCE WITH THE MEASURES OF THE METRIC SYSTEM.

By degrees our Federal government may introduce
the metric weights and measures into our public surveys ; such as the coast survey, the several boundary
surveys, the geological, topographical, and land surveys of the territories, and the surveys of the lakes.
In the published reports of these works, or at least in
such of them as are intended for, or are likely to have,
a large circulation among the people, it would be
advantageous, and would familarize metric values to
the popular apprehension, if dimensions, quantities,
and weights should be expressed both in metric denominations and those of the existing system.

NAVAL AND MILITARY ESTABLISHMENTS REQUIRED TO USE THE SYSTEM.

The metric weights and measures may further be
introduced into actual use in the navy yards and
military posts maintained by the government in the
different parts of our territory ; and, finally, the busi-

ness of the post office department may be largely, if not wholly, conducted, so far as weights and measures are concerned, in metric denominations.

THESE MEASURES RECOMMENDED BY THE PARIS CONFERENCE OF 1867.

These are measures which were unanimously recommended by the international conference on weights, measures, and moneys, which was convened in Paris, in 1867, consisting of delegates appointed by the governments of twenty-two different nations, including, of those not using the metric system, Austria, Russia, Sweden, Norway, Denmark, Turkey, Great Britain, and the United States. To most of them, as it appears to me, there can be no reasonable objection, even on the part of those who have no admiration for the metric system themselves, and no faith in the prediction of its final prevalence. If nothing follows them, they can at least do no harm.

CONCLUSION.

I have occupied, gentlemen, a larger portion of your time than I intended, and larger, I fear, than will have seemed to you reasonable. The subject itself is a large one, and my interest in it is deep. I am so far from pretending to have exhausted it, that I feel that what I have said is but the merest skeleton of an argument. I wish to be indulged only in a single additional remark, which shall be in regard to the able and comprehensive, and, at times, eloquent report of Mr. JOHN QUINCY ADAMS, which you have republished in the same volume with the report of your committee.

The original publication of that report, able and powerful as it is, and for the very reason that it is able and powerful, I esteem to have been a serious public misfortune. It effectually extinguished all hope of metrological reform in the United States for half a century. And yet Mr. ADAMS, decidedly as he discouraged any legislation, at least for the time being, and apparently for a very long time, looking toward the recognition of the metre in America ; darkly as he drew in the lines as he painted the picture of France writhing in the toils which the metric system had thrown round her ; and fondly as he lingered over that beautiful system of British weights and measures distinguished by the priceless property of a " uniformity of proportion " of which he laments that there remain to us only the ruins ; Mr. ADAMS, after all, was an admirer of the metric system to such an extent, that one is sometimes at a loss to decide whether he seems to love or to fear it most. In the midst of his doubts and his misgivings, he cannot refrain from occasionally enlarging upon its merits, in language strong enough to satisfy even the most enthusiastic of its advocates. And when for a moment he succeeds in forgetting France, and in shaking himself free from the embarrassing associations of the immediate present, he becomes

as it were inspired with a spirit of prophecy, under
the influence of which he is oblivious of all difficulties, and glowingly anticipates that very approaching triumph which his own labors are destined so considerably to postpone. No words that I can use can
add to the positiveness of assertion with which he pre-
dicts that final consummation to which I have declared
to you to-day that I so confidently look forward. I
cannot do better, therefore, in concluding these remarks, to which I thank you for having so indulgently
listened, than to adopt his own language, and to express with him the conviction that, "If man upon
earth be an improvable being, if that universal peace
which was the object of a Saviour's mission, which is
the desire of the philosopher, the longing of the phil-
anthropist, the trembling hope of the Christian, is a
blessing to which the futurity of mortal man has a
claim of more than mortal promise ; if the Spirit of
Evil is, before the final consummation of things, to be
cast down from his dominion over men, and bound in
the chains of a thousand years, the foretaste here of
man's eternal felicity ; then this system of common
instruments to accomplish all the changes of social and
friendly commerce will furnish the links of sympathy
between the inhabitants of the most distant regions ;
the metre will surround the world in use as well as in
multiplied extension ; and one language of weights and
measures will be spoken from the equator to the
poles."

APPENDICES.

APPENDIX A.

NOTE ON THE UNIFICATION OF MONEYS.

(REFERRED TO ON PAGE 24.)

THE expediency, or rather, in fact the necessity, of treating independently and separately the difficult question of the unification of the monetary systems of the world, is made immediately evident when we consider that, though moneys are measures of value, their relations to the measures of material quantity are not fixed by any necessary law. In all systems of ordinary metrology, to settle the standard of length, settles the standards of capacity and weight as well; but it leaves the standard of money still unsettled, and permits most widely different views to be maintained on this point, with almost equal plausibility. There are besides, as mentioned in the foot note on page twenty-four, a number of considerations which specially complicate the monetary problem; and to postpone the simpler question of metrological reform until all these can be disposed of to the satisfaction of all nations, would be as unwise as it is unnecessary.

The popular impression is undoubtedly quite the reverse of this. It is an opinion easily taken up upon a *primâ facie* view of the subject, that the pound sterling of Great Britain, the dollar of the United States, and the franc of the French republic, might, by the simplest process of legislation in the world, be reduced to a regular geometrical series of values. Hence therefore it is probable that everybody, who knows anything about these important national money-units and the relations

in which they stand to each other, has often said to himself, in
the language of one of the resolutions reported to the Convocation of the University of the State of New York, by their committee on weights, measures, and coinage, that "such changes
should be made in the values of the franc, the dollar, and the
English pound sterling, that five francs be exactly equal in value
to one dollar, and five dollars exactly equal in value to one
pound sterling." It was surely not an observation so little profound as this, that the convocation had a right to expect of
their committee. Considered as a help toward the attainment
of the desired uniformity of currencies, it rather reminds one of
good Mrs. Nickleby's very practical suggestions to her embarrassed husband. "My dear Nicholas," said this excellent lady,
"why don't you *do* something? Why don't you make some
arrangement?"

The committee propose to the convocation to resolve that the
discordance existing between the monetary systems of three great
nations ought to disappear. But this is only what all the world
resolved long ago ; and therefore it has happened, that the sun-
dry international conferences which have been called during the
last fifteen years to consider this subject, have wasted no breath
in axiomatic propositions and empty generalities, but have given
all their collected strength to the study of the knotty problem,
how this thing is to be done. To the solution of this problem,
considered by the ablest publicists and profoundest financiers
of all nations to be one of the most perplexing with which legislation has to grapple, the committee might surely have contributed something, had it been only in the way of making known
the difficulties by which the question is embarrassed. This
would have been better than, like Mrs. Nickleby, to recommend that the nations should "make some arrangement."

Without attempting to exhaust the subject, let us hint at a
few of these embarrassments. The pound sterling, the dollar,
and the franc are respectively units of the money of account

of the peoples of Great Britain, the United States, and France. The pound sterling is represented in currency by a gold coin, called a sovereign; the dollar, by a coin of either gold or silver, called also a dollar; and the franc, by a silver coin, bearing likewise the name franc ; but a coin of five francs is struck either in gold or in silver. The British mint coins silver, but gold is the standard metal, and the silver coins are legal tenders only for small amounts. In France and the United States, gold and silver are both standard metals—or there exists legally what is called a double standard; hence all gold coins in either country, and all silver dollars in the United States, or silver five franc pieces in France, are legal tenders for all amounts. The theoretic basis of the French system of coinage is the silver franc (composed of nine parts of pure silver to one part of alloy), having the metric weight of five grammes. This coin, which is now no longer issued from the mint, and which has long since ceased to have a practical existence, was also a legal tender for any sums while it lasted. The present silver currency of France, with the exception of the silver five-franc piece, is a debased coinage, and is not legal tender except in small transactions. Similarly, the silver currency of the United States below the silver dollar, is, if not debased, at least degraded, by being diminished in weight, so that its nominal value is greater than its real value, and it is not a legal tender except for trifling sums. Now the reason why the silver legal tenders have disappeared from circulation in France and the United States, is explained as follows.

Since, to coins of determinate legal value, determinate weights must be assigned and permanently maintained, it follows that, where the double standard exists, the *relative* value of gold and silver must be fixed by law. But the relative value of these two metals is not fixed in the nature of things, any more than is that of any other two commodities. It varies with the varying supplies of these two metals respectively, and of the varying demand for them, not only for the purposes of coinage, but also for the uses of luxury and of the arts. Accordingly, in the course of the last few centuries, it has widely fluctuated in Europe. About the middle of the thirteenth century, gold was worth in England only about ten times as much as the same weight of silver. A century later, the ratio stood as twelve and a half to one. The highest point it has ever attained appears, by the researches of E. B. ELLIOTT, Esq., Statistician to the United States Treasury Department, to have been fifteen and eighty-three one-hundredths to one. The same authority puts this ratio, now, at fifteen and thirty-eight one-hundredths to one. Now, the ratio between legal tender silver and legal tender gold is, in the United States, sixteen to one. And as a gold dollar will pay a debt as well as a silver dollar, while the gold dollar is really worth only fifteen and thirty-eight one-hundredths times its own weight of silver, and the silver dollar weighs *sixteen* times as much as the gold dollar, it is evidently for the interest of the holder of silver dollars to deprive them of their character as coins, by melting them up ; and to sell the resulting bullion for gold dollars. Hence the disappearance of our legal tender silver, and the degradation by law of our smaller silver currency to a degree which brings down the ratio so low as to fourteen and eighty-eight one-hundredths to one.

The French have fixed the ratio between these two metals at fifteen and five-tenths to one. This is nearer the present truth, but it is still sufficiently too high to allow the circulation of legal tender silver ; and, accordingly, the legal tender silver franc, which is the theoretic basis of the system (in the absence of which this system, considered as a metrical system, is completely fictitious), has ceased to be coined, and exists only in name.

Now, if the law in France which fixes the relative value of the two standard metals should be slightly changed, so as to make the ratio fifteen to one, it is obvious that the silver five-franc pieces, of which many still exist, would immediately return to circulation, and would be in great demand; also, that the legal tender franc might again be safely coined, though possibly it might not be coined largely on account of the greater wear of small coins by abrasion. And from this time forward it is equally obvious that gold would begin to disappear.

The impossibility which is thus seen to exist, of maintaining two legal tender coinages of different metals side by side, would seem to most minds, at the first examination of the question, to be a quite conclusive reason against the adoption of the double standard. And yet it is on this very fact that some of the most eminent publicists of France have founded their strongest arguments in defence of a system so seemingly vicious. At the International Conference which was called in 1867, to consider questions of money, weights and measures, Mr. WOLOWSKI urged that, when the legal ratio is fixed nearly at the mean actual ratio, the tendency, in case of fluctuations, to bring at one time one metal into demand, and at another time another, exerts a steadying effect upon values, and operates to retard these fluctuations, and to diminish their extent.

Now there is solid reason in the view thus taken ; and if the tendency spoken of were to be left undisturbed, it would certainly temper, if not entirely obviate, the undeniable evils attendant on the double standard. But, historically, it is certain that this tendency has never been left undisturbed; and there is therefore good reason to believe that it never will be. The fact has constantly been, that, so soon as one of the metals has begun to be crowded out of circulation, the attempt has been made to legislate it back by degrading or debasing the coinage in that metal. And thus the maintenance of the double standard, instead of maintaining the equilibrium of ratios, as imagined by Mr. WOLOWSKI, operates as a constant temptation to the commission of the gravest of all possible errors of legislation in matters of finance.

But, obviously, we are in no condition to agree upon a common system of money for all the world, until we can, at least, agree first upon a common representative standard of money values. The French begin with a silver franc of determinate fineness and simple determinate metric weight. From this, by a numerical relation of values, entirely arbitrary and artificial, they deduce the weight of the one-franc theoretical, or of the five-franc piece actual, of gold of similar fineness ; which weight is fifty thirty-first parts of one gramme—a quantity no longer simply metrical, nor expressible by a finite decimal. But this five-franc piece of gold, which is not metrical in weight, the Paris Conference of 1867 recommended to the world as the basis of an international coinage ; making this recommendation in the same breath in which they recommended the universal adoption of · the metric system of weights and measures.

It is manifest that the double standard cannot be maintained. But supposing the double standard abandoned, it does not follow that the single standard of *gold* will be immediately and universally accepted. Of nations recognizing but a single standard, some, like Holland, prefer silver to gold. It is notorious that the Oriental peoples, who have absorbed the silver dollar coinage of Spain, Mexico, and the United States, as fast as it could be produced, have always steadily refused the sovereigns and the eagles and the doubloons and the napoleons. Before we can proceed a step, therefore, toward the creation of an international monetary system, we must not only agree upon a single standard, but agree also what that standard shall be.

But such an agreement is by no means so simple a matter to reach as it seems. Notwithstanding that the drift of opinion among statesmen and authorities on finance is decidedly towards gold, the need of a coin which will be received by the people of China and Japan is one which must be met ; while the metric nations of France, Belgium, Italy, and Switzerland, who are allied for the maintenance of a common monetary system, will

abandon very reluctantly, if at all, the alternative silver standard; since, in abolishing that, they must abolish at the same
time the agreeable fiction which maintains their coinage in
decimal relations with the metric system of weights and
measures—one ideal napoleon of silver weighing exactly one
hundred grammes, while one actual napoleon of gold weighs
6.4516129 + grammes, the decimal being without end.

But, supposing this difficulty to be somehow or other disposed
of, we presently encounter another, usually quite overlooked in
popular or conversational discussions of this subject—a difficulty
which, nevertheless, is grave enough to require no small effort to
remove it. It is well known that the precious metals, when in a
state of purity, are too soft to be advantageously employed as
the material for coinage. Their alloys offer greater resistance
to abrasion, and therefore wear longer. On this account such
alloys are preferred, for the uses of the mint, to the pure metals.
But as the weight of pure metal (gold or silver) in a coin determines its value, so the total weight of any given coin must be
increased in proportion as the metal of which it is composed is
more largely alloyed. It is therefore obviously advisable to
introduce as small an amount of base metal into the compound
as practicable, consistently with the attainment of the desired
hardness. But the question, what are the proportions which
under a given value combine the maximum of durability with
the minimum of weight, is one which, as a question of pure
science, has not yet been solved. Accordingly, the statutory
provisions established by different governments in regard to the
standard fineness of their coins, are widely discordant. For
coins of gold, the proportion of nine parts of the precious metal
to one part of alloy has been adopted by our own country, and
by France, Belgium, Italy, Switzerland, Prussia, Bavaria, and
Spain. Austria employs the same for her gold crown and half-
crown, and Holland for her double-William and its sub-multiples. But for her ducat and double-ducat, Holland employs an

alloy containing fifty-nine parts of pure gold to one of alloy ; and Austria, for a coin of the same name, prescribes a fineness of seventy-one parts of pure gold out of seventy-two of total weight.   This latter is also the proportion employed in the gold coins of Wurtemberg.

The standard of fineness in use in Great Britain, Portugal, Brazil, and Turkey, consists of eleven parts of pure gold to one of base metal ; and that of Sweden, of thirty-nine parts of pure gold to one of the alloy.   In other countries the standard falls below nine-tenths.   In Denmark it is eight hundred and ninety-five one-thousandths ; in Russia, eighty-eight one-hundredths ; in Mexico it is different in different coins, being eight hundred and seventy-five one-thousandths in the piece of twenty pesos, and eight hundred and sixty-six one-thousandths in the doubloon ; in Bolivia it is eighty-seven one-hundredths ; in Ecuador, eight hundred and forty-four one thousandths; and so on, reaching in the Japanese cobang the extreme debasement represented by five hundred and seventy-two parts of pure gold to four hundred and twenty-eight of alloying metal.

Similar diversity of legislation is observable in regard to silver coins ; but as to these there is a more general conformity to the standard of nine-tenths.   Such is the fineness of the large silver coins of the United States, and of France, Belgium, Italy, Switzerland, Prussia, Bavaria, Wurtemberg, Baden, Hesse Darmstadt, Austria, and Spain ; while England employs an alloy of thirty-seven fortieths; and Holland, one of nine hundred and forty-five one-thousandths.   In other countries the standard is generally lower; and in most or all of those above named, the subsidiary silver coinage is materially less fine.

As the several German States named in the foregoing paragraphs are now members of the German empire, the diversities noticed among them will probably soon disappear.   There appears to be everywhere a growing disposition to adopt the uniform standard of nine-tenths.   But to change a standard of this kind is by no means so simple a thing as it seems. Suppose, for instance, there happens to be already a vast existing coinage, circulating among a people who propose to make such a change. The modification of the standard changes the appearance of the coin to the eye (the color, in the case of gold) and changes also the weight, without, nevertheless, changing the value. Both these changes are disadvantageous to the freedom of circulation. In our own country, for example, differences of color between different coinages of eagles and double-eagles, arising out of the fact that, in the refining of the native gold, the metals naturally associated with this have not always been separated with equal thoroughness (though the residual impurity has always been carefully determined and allowed for as part of the alloy), have been regarded by the director of our mint, Gov. POLLOCK, to be sufficiently prejudicial to the character of the coin to call for notice in his communications to the Treasury Department, and to require measures to be taken for their correction. But the difference in the *weight* of coins produced by varying the proportions of the alloy, is the occasion of a much more serious disadvantage. To count out coined money in large sums is a very time-consuming operation. In minting establishments and in great banks of deposit, there may be, and in the former case there usually will be, found mechanical contrivances for counting by hundreds of pieces at once (available, however, only when the coins are all of the same denomination); but it is not to be supposed that tradesmen, merchants, or even bankers generally can be provided with such conveniences. Coins, when transferred from hand to hand in large masses, are therefore rarely counted. They are simply weighed in bulk ; and the weight, when the composition is uniform, *is* an infallible indication of the value. But this cannot be the case if the mass consist of mixed coins, some of them of one degree of fineness and some of another. Hence, whenever the standard of fineness is changed by law, it becomes desirable and even almost necessary, to withdraw from circulation all the legal tender coinage in the same metal already in existence. This consideration is one of no trifling importance to a people who, like the British people, have no less than eighty millions of pounds sterling—say four hundred millions of dollars—in the form of gold sovereigns and half sovereigns. This is the estimate of the existing British gold coinage, as made by the eminent statistician, Prof. JEVONS, in a communication made in 1868 to the Statistical Society of London, and corroborated by Mr. MILLER, of the Bank of England, who is quoted as authority to this effect by Dr. FARR, delegate from Great Britain to the International Statistical Congress, held at the Hague, in 1869, in a report made by him to that body. It may easily be conceived that Great Britain would bring herself reluctantly to the adoption of the standard of nine-tenths, even though convinced that it is a better standard than that which she employs at present. Her delegates to the International Conference of 1867 gave their assent to the proposition favoring the nine-tenths standard, though not concealing their preference for their own,* solely, as it would appear, because they saw no hope of a general agreement upon any other.

But let us suppose that we have got over the difficulty of the standard metal, and of the double or single standard, and also over any difficulty as to the standard of fineness; we find yet behind a difficulty far more troublesome. The visible representative of our unit of account is to be, we will say, a coin of gold of the fineness of nine-tenths; the question now arises, what shall be the *weight* of this coin? It is easy to see that this question may be so answered as to demonetize at once the entire metallic currency of the world; and that, if we are ever to have an universally accepted international coinage, it *must* be

---

* Report of the Master of the Mint, Sir THOMAS GRAHAM, and Mr. C. RIVERS WILSON, delegates to the International Monetary Conference, made Dec. 2, 1867, to the Lords Commissioners of the Treasury.

so answered as to produce, to some extent, and, it may be, to a large extent, an effect of this kind. To touch an interest of so vast magnitude as the coinage of the world, is a thing which, to be done safely, must be done cautiously. It is a low estimate, to place the value of the gold and silver coin now existing at twenty-five hundred millions of dollars. This is the estimate of Mr. McCulloch, in the article on the "Precious Metals," contributed by him to the *Encyclopedia Britannica*. But in this he attributes to Great Britain only from seventy to seventy-five millions sterling of coin in both gold and silver, taken together; while Professor Jevons and Mr. Miller, as we have seen, place the gold alone as high as eighty millions, while putting the silver coinage at fourteen millions: thus making a total of ninety-four millions.

The coinage of France Mr. McCulloch puts at from one hundred and thirty to one hundred and forty millions sterling. That of the rest of Europe, of North and South America, of Australia, of the Cape of Good Hope, and of Algeria united, he supposes may equal three hundred millions ; so that, summing up all these several estimates, he would make out a grand total for the world, excluding Asia and all Africa excepting Algeria and the Cape, falling somewhere between four hundred and ninety and five hundred and ten millions—say in round numbers five hundred millions, which is the mean of the two, and is nearly equivalent to twenty-five hundred millions of dollars, as above.

From statements made by the Hon. Samuel B. Ruggles, the distinguished representative of the United States at the International Monetary Conference of 1867, in a written argument addressed to the conference, we should be justified in carrying the total much higher, perhaps to not less than three thousand millions of dollars. Mr. Ruggles stated that the mint of the United States had coined, up to that time, eight hundred and forty-five millions of dollars, in all ; and that, of this sum, probably three hundred millions still remained in the country. He took the

estimates of DE PARIEU for France, Belgium, and Italy, at fourteen hundred millions; and allowing four hundred millions more
for the remaining countries of continental Europe, he made the
total for the continent entire, eighteen hundred millions. This
last assumption is considerably more than justified by the estimates of McCULLOCH, since the latter authority puts down Russia
alone for two hundred and seventy millions of dollars ; which
would leave, comparing this statement with that of Mr. RUGGLES,
only one hundred and thirty millions for Spain, Portugal, Holland, Denmark, Sweden, Norway, Switzerland, Austria, Turkey,
and all Germany.

Taking, however, the eighteen hundred millions of this report
of Mr. RUGGLES as the total for continental Europe, and his three
hundred millions as a proper estimate for the United States, with
the four hundred millions of McCULLOCH for Great Britain, we
shall have the twenty-five hundred millions complete, without
going any further, or making any allowance for Mexico, British
America, the West India Islands, Central America, South America, Australia, the Cape, Algeria, or Egypt. Considering that
we have, among these, some of the principal gold producing and
silver producing states of the world, it is a small estimate to
suppose that they have among them the five hundred millions
necessary to bring up this result to the round total of three
thousand millions, named above. It must further be noted that
no account is taken in either of these estimates of British India,
or of China and Japan, which latter countries have been large
recipients of the silver coin of Europe and America.

A more recent and very carefully considered estimate by Mr.
RUGGLES, embraced in his supplementary report, as delegate to
the International Monetary Conference above mentioned, made
April 8, 1870, to the Secretary of State of the United States,
reduces this total to twenty-five hundred and fifty millions of
dollars. As this estimate classifies the nations in groups as
using the gold franc, the gold dollar, the gold sovereign, or

other coins of gold, it has an important practical interest in connection with the question of the unification of coinage. Mr. RUGGLES says :

"According to the estimates of political economists and bankers in Europe and the United States, believed to be reliable, the amount of gold coin now outstanding, in the various parts of Europe and America, is in round numbers :*

"I. In France, Belgium, Switzerland, Italy and Pontifical States, using the gold franc $1,250,000,000

"II. In Austria, Spain, Sweden, Greece and Roumania, agreeing to use the franc..... 200,000,000

$1,450,000,000

"III. In the United Kingdom using the sovereign................... 450,000,000

"VI. In the United States using the gold dollar............................... 200,000,000

"V. In Germany, (North and South,) Netherlands, Denmark, Norway, Portugal, Russia and Turkey using other gold coins, all differing in value..................... 300,000,000

$2,400,000,000

"VI. In Canada, Mexico, Central America and South America using sovereigns, dollars, doubloons and various other gold coins.. 150,000,000

$2,550,000,000

---

* To make the exhibit complete, the following statement of the population and foreign commerce of the several groups of nations above given, should be added. It is derived also from the report of Mr. RUGGLES referred to.

	Population.	Foreign Commerce.†	Gold Coin Outstanding.
Using the franc.............135,700,000		$1,648,000,000	$1,250,000,000
" pound............ 30,400,000		1,251,500,000	450,000,000
" dollar............ 38,555,983		438,500,000	200,000,000
" other gold coins...135,500,000		852,000,000	300,000,000
" various coins..... 373,600,000		240,000,000	150,000,000

† Deducting duplications: in the full statistics of commerce, the same values appear twice.

It is evident from the foregoing, that great exactness is not yet attainable in the statistics of this subject. What can only be exactly known are the issues of the several mints. The history of those issues after they have passed into private hands, is not easily traced. Dr. FARR, in his valuable and elaborate report, above spoken of, to the Statistical Congress at the Hague, gives the methods and results of some interesting investigations concerning the average life of the British gold sovereign. Such inquiries throw much light upon the subject ; but as the conditions of the life of coin are not necessarily the same in all countries, they must be prosecuted longer and more generally before they can be regarded as having acquired for this branch of statistics the character of an exact science.

Yet, whatever may be the degree of confidence or of doubt with which we speak in definite terms of the immense values actually existing in the world in the shape of coin, we can entertain no doubt at all that these values are really immense. And hence we perceive the serious gravity of the interests liable to be affected by the answer which may be given to the question, what shall be the weight of the visible representative of the international monetary unit? The persons who settle great questions of international coinage over the breakfast-table, or block out schemes for the reconciliation of discordant monetary systems in their moments of leisure or recreation, frequently express impatience of the sluggishness with which the world advances in a path so plain ; and account, perhaps, for the phenomenon, on theories of the indifference or unskilfulness of diplomatists, or of the obstinacy of ministers. It rarely occurs to such persons to consider that diplomatists and ministers have, in fact, as great an interest in the happy solution of this important question as they; and are, perhaps, making progress as rapidly as they themselves would be likely to do, if they had the management of the whole business in their own hands. The grave fact, which has rendered unavailing the labors of all the international monetary conferences of the last fifteen years, has been, that no common currency can be adopted by the nations, without condemning the entire coinage of one or another of them, and perhaps of several, to the melting pot. This consideration, and the consequences it involves affecting the liquidation of existing debts in coin of altered weight, determined the State Department of our own country to propose to the governments of Europe, about a year ago, a plan for giving to the gold coin of all nations an international circulation, without essentially altering their present actual weights or values. The plan here referred to consists in assuming a common unit of value so small, that the most important gold coins of all nations may be made, by slight changes, exact multiples of this. And in order to avoid any embarrassment from diversity in the standards of fineness, it also proposed to make the unit a definite weight, not of alloyed gold, but of *pure gold*. The unit weight proposed was one decigramme. The value of one decigramme of pure gold is very nearly six and two-thirds cents. The gold dollar contains of such pure metal fifteen decigrammes and a little more than forty-six one-thousandths of a decigramme. By dropping the small fraction in excess of fifteen decigrammes, the value of which is but thirty-one one-hundredths of a cent, the dollar becomes just fifteen times the unit ; and the half-eagle, seventy-five times the unit. The sovereign, by a similar slight adjustment, becomes equal to seventy-three times the unit ; the napoleon, to fifty-eight times the unit ; the double ducat of Austria, to sixty-nine times the unit ; and so on with respect to other coins.

The merit of this scheme, and it is not a small one, is, that it would bring the monetary units of the different nations into relations of commensurability, without impairing the usefulness for the purposes of money of the coin actually in existence. The disadvantages attending it are, that the ratios of commensurability which it establishes are lacking in the simplicity which is the indispensable condition of practical usefulness ; and that the diversity of alloys employed in the coins of the different nations must always prevent any national coinage from securing a free international circulation. This scheme, therefore, though simple and at first view seemingly feasible, does not appear, upon mature consideration, to offer an adequate solution of the difficult problem before us.

Other propositions, looking to the same end, have been :

1. To adopt as a unit of account the value of some convenient coin now actually existing, with its corresponding weight in gold ; a proposition which is equivalent to requiring that all the systems of coinage now in use, except one, shall be abandoned ; and that this one, from being simply national, shall become international and universal.

2. To adopt a value represented by no coin at present, but of which the representative in gold shall be the unit base of such system of weights as may be adopted for international use (say the metric system), or shall bear a decimal relation to that base.

3. To adopt a value of which the representative in gold shall be in simple, though not necessarily in decimal, relation to the base unit of the system of weights ; and which shall itself be such that the coinage of the principal commercial nations may be conformed to it without very large changes of weight.

In considering the first of these propositions, no existing coins will be likely to suggest themselves as possessing claims worth attention, except the gold piece of five francs, the gold dollar, or the gold sovereign. The sovereign, out of England, is likely to find no support ; being elsewhere regarded generally as representing too large a value to be conveniently employed as a unit of account. In fact, the British commissioners to the international conferences have themselves proposed, rather, a determinate part of the pound sterling—as an eight-shilling piece, for example—than the pound sterling itself. The piece of five francs, with its present actual weight, and actual fineness of nine-tenths, was recommended as the international unit by the monetary conference of all nations, held at Paris in 1867; and it is now warmly advocated by men eminent as statesmen, statisticians, and publicists, in the United States and on the continent of Europe. The representatives of France in that conference went even so far, in the ardor of their desire to win over other nations to a measure so flattering to their country, as to consent to the abolition of the double standard and the adoption of the gold standard only ; although by so doing they would be compelled to sever the connection between their monetary system and their system of weights and measures ; thus presenting to the world the remarkable spectacle of the first of metric nations recommending and advocating a non-metric coinage.

It is not, however, to any sentiment of a merely selfish nature, that this important concession on the part of these gentlemen is to be ascribed. The result must rather be regarded as a reluctant sacrifice of a cherished ideal, to what seemed to them, at least, the constraining force of accomplished facts. Some idea of the weight of the considerations which led them to their conclusion, may be gathered from the following summary of the debate in the conference, given by Mr. RUGGLES in his supplementary report of April 8, 1870.

"The debates were systematically conducted by means of questions methodically and carefully arranged. They were commenced by the discussion much at length of the following preliminary and comprehensive question :

" ' By what means is it most easy to realize monetary unification : whether by the creation of a system altogether new and independent of existing systems ; or by the mutual coördination of existing systems, taking into account the scientific advantages of certain types and the number of the populations which have already adopted them ?'

" On this cardinal question, the debate was prolonged and important, in which the delegates from many of the nations took part. It was commenced by M. MEES, delegate from the Netherlands and President of the Netherlands Bank, who 'declared that if he could admit the immediate realization of the unification of coinage, he would give the preference to the first of the two alternatives. In this case, in effect, the creation of a new system, avoiding all national susceptibilities, would seem to him the best way to obtain the end. But it does not seem possible to him, that complete uniformity can be speedily attained, and therefore he considers the second alternative as being alone of a nature to produce actual practical results.'

"The Count D'AVILA, delegate from Portugal, and recently its Minister of Finance, maintained that 'if the different States found themselves obliged by the establishment of a system altogether novel to change simultaneously their monetary regulations, the difficulties of the attempt would be multiplied in such manner that they would become insurmountable.' He further deemed it 'essential that an agreement should take place between England, France, and the United States,' that it 'would sooner or later rally also the other countries,' and that the 'example would have a decisive effect.' He declared himself ready to vote for a single gold standard, the reduction of the pound sterling to twenty-five francs, and the American dollar to five francs, with the gold coin of five francs as the monetary unit.

"The delegate from Austria, the late Baron DE HOCK, favorably known by his works on finance, concurred in the opinion expressed as to the impossibility of securing the acceptance of an entirely new system, and completely breaking up inveterate habits. 'In Germany,' said he, 'we find a striking example : there was a wish to introduce into the German States a coin not correspondent with any existing types. Although it was the most rational and accorded *perfectly with the metric system*, it could not find its way into calculations. The *gold crown* only passed from the mint into the melting pots of the goldsmiths.

It is only as expressed in the second sentence of the first question, by the mutual coördination of existing legislation, by taking into account the scientific advantages of certain types, and the number of the populations which have already adopted them, that a solution may be found.' He added that gold, 'which has spread in such considerable amounts through the European markets during the last twenty years, would be the most convenient agent for a universal monetary circulation.'

"M. FEER HERZOG, one of the two delegates from Switzerland and familiar with its financial affairs, said that 'there is in France a school important because of the scientific authority of its adepts, which admits no other monetary unity than metric unity in round numbers, and proposes to take for the unit a weight of five grams of gold, nine-tenths fine,' (equivalent to $2.99\frac{0.9}{100}$ of the present money of the United States). 'This theoretic solution would be wanting in one essential quality, that of practicability. At the time we have arrived at, we cannot invent a monetary unit not in relation with any type actually existing. The franc itself had been compelled not to depart too far from the *livre tournois*, in order to make itself acceptable, and the gold crown, containing ten grams, has not been able to get into circulation in Germany because it is not adapted to the florin of Austria, nor the florin of Bavaria, nor the thaler of Prussia. By the very force of things,' said he, 'it is the napoleon [twenty francs], a foreign coin, which represents beyond all others the monetary circulation of gold in Germany.'

"The delegate from Russia, M. DE JACOBI, Privy Councillor of the Crown, and member of the Imperial Academy of Sciences at St. Petersburg, highly distinguished in Europe not only in general physical science, but especially by his earnest and able advocacy of the merits of the metrical system, said that he 'adopted in full the ideas developed by M. FEER HERZOG. He would have been glad that a relation should exist between coins and the systems of weights and measures, but in the double view of science and practice, he saw no necessity for the establishing of such relations to the prejudice of other more important interests.   He could not, therefore, regard as serious the reproach cast upon the coin of France, as having widened the breach in the French metric system.' He added that 'the creation of an entirely new coinage was so much the less opportune, that he could not let the occasion pass without noticing the agreement, perhaps accidental but almost complete, between the intrinsic value of the principal French coins and those of Russia.   Thus the silver rouble coincides very nearly with four francs, the difference not exceeding the limits of the tolerance. In the same way the demi-imperial has a value only fifteen kopecks ' (about eleven cents,) ' higher than the twenty franc gold piece.

"M. MEINECKE, the delegate from Prussia, thought it of prime necessity to adopt as the base of the new system, a system already recognized and reduced to practice. ' The difficulty,' said he, ' of adopting the gold standard, is much greater in Prussia than for any other country; but if the labors of the conference should aim at establishing the basis for a general monetary arrangement, Prussia would study with care the best mode of connecting herself with it.'

" On the other hand, Mr. STAS, one of the two delegates from Belgium, said that he ' would prefer the establishment of an entirely new monetary system, and that the conference assume as its mission, to settle principles and not expedients in practice. To follow the latter course would be to leave traces in the snow, not to engrave footprints in rock.' He also maintained that ' the creation of a system based on a standard of gold of five or ten grams would offer the immense advantage of having it more readily accepted by all nations, as it would avoid all national susceptibility.   Doubtless the adoption of the new unit would require the general reminting of all coinage, but this recoinage would bring with it a definitive system sanctioned by science.'

He further asserted that, 'mathematically speaking, the kilogram cannot be divided into one hundred and fifty-five equal portions.'

"Mr. Feer Herzog, replying to the remarks of Mr. Stas, said : 'there is nothing to hinder the definition of this napoleon by indicating the round number one hundred and fifty-five, which a kilogram includes, or rather it should be divided by the fractional number of grams which represent its weight, neglecting the decimals beyond the thousandths decimals, which practically are of no importance and have only an interest purely scientific. It is not indispensable to the goodness of coin that it should be metrically proportioned.'

"The delegate of the U. S. [Mr. Ruggles] said it would be as impossible to abolish ' the expression of the dollar in the United States, as that of the sovereign in England, but that both might be retained in reducing their intrinsic values.' He said that ' two milliards ($2,000,000,000) in gold had been thrown into the money market since the discovery of the mines of Australia and California, and that it was certainly possible that the coinage of gold in the United States, in the next fifteen years may reach five milliards of francs ($1,000,000,000). In view of such a future, the American government would prefer to reduce its monetary unit at once.* But the United States in consenting to recoin its gold now in circulation, would expect France on her side, will consent to coin pieces of twenty-five francs, in which case monetary unification would at once assume a practical form.' "

"On closing the debate the question was taken by ayes and noes, on a roll-call of the nations, which resulted in an unanimous vote in favor of the second alternative in the question, in the following words, to wit :

"'It is more easy to realize monetary unification by mutual coördination of existing systems, taking into account the scientific advantages of certain types, and the numbers of the populations which have already adopted them.' "

* Approval of Department of State, in letter from Mr. Seward, June 21, 1867.

Subsequently, also, in a debate in the French Senate, on the expediency of immediately issuing, in furtherance of the plans of the conference, a gold coin of the value of twenty-five francs, the conflict between the perfection of theory and the possibilities of practice was brought very sharply out. The following is an abridged outline of the discussion, as given by Mr. RUGGLES.

"In this interesting and instructive debate, which occupies fourteen printed columns of the *Journal Officiel*, the comparative merits of the twenty-five franc coin and of a new coin of ten even grams (or a decagram), equivalent to thirty-one existing francs, were fully examined. The metrical merits of the proposed decagram coin had been urged on the twenty-first of January, by Mr. MICHEL CHEVALIER, in a general speech on the coinage of France, in which he had claimed for the decagram its perfect conformity to the 'logic of the metrical system.

"It is, however, specially noticeable that in the latter portion of his speech, Mr. CHEVALIER, who combines with his abstract philosophies, a large admixture of practical statesmanship, after stating these 'indications of logic,' very pertinently proceeds to ask and then to answer as follows :

"' Is it necessary to abandon ourselves absolutely and exclusively to logic ? It is a question. Logic is a power ; but various conventional requirements *(convenances diverses)* may cause us to disregard it *(peuvent faire qu'on s'en écarte)*.

"' When, in order to conform wholly to logic, it is necessary to submit to considerable sacrifices, or to change the inveterate and cherished habits of peoples : and when even the triumph of logic is not to be followed by great advantages, we are allowed to hesitate. A wise administration may lay aside logic, and adopt combinations which from other points of view would be advantageous to the public interest.'

"The debate of the twenty-fifth of January was opened by Mr. LE VERRIER, the astronomer and *savan* measuring the outmost limits of planetary space by his skilful use of exact science. In a powerful speech of wide range, he supported the plan of the twenty-five franc coin, as being practicable, in preference to that of the decagram, as being merely theoretical. He was followed by Mr. DUMAS, 'President of the Commission on Coins,' and as such, Director of the Mint, who stated that the recoinage required by the decagram, would cost seventy millions of francs, or fourteen millions of dollars. Mr. LE ROY DE SAINT-ARNAUD concurred with Mr. LE VERRIER, and held that 'the question could be resolved only in accordance with practical ideas. I admire science,' said he; 'I bow before the *savans ;* but in questions of this nature, in which so many persons and interests are intermingled,' * * * 'I regret the excess of science, and fasten myself to the facts.' He was followed by General Marquis DE LAPLACE, who denounced the twenty-five franc coin as unmetrical and unnecessary, and urged among other objections, that it differed too little in size from the twenty-franc coin, and might tend to raise to twenty-five francs, the price of objects now selling at twenty.

"In closing the debate, the necessity for the immediate issue of the twenty-five franc coin, and indeed the whole plan of the Paris conference, were vindicated with signal ability by Mr. DE PARIEU, the universally acknowledged leader in Europe of the pending measure of monetary unification, and now president of the council of state in the new liberal ministry. Suffice it to say that in accordance with the form of proceeding peculiar to the body, the Senate 'passed to the order of the day,' thereby rejecting the proposition of the decagram, and virtually expressed their approbation of the twenty-five franc coin by leaving its issue to the proper minister charged with the coinage."

Since 1867 some progress has been made on the continent of Europe toward unification on the plan here proposed. The following further extract from the supplementary report of Mr. RUGGLES shows to what extent this is true:

"Up to the present time the recommendation of the conference in respect to the five franc gold unit (deferring the immediate consideration of the question of the single standard), has been adopted by nine of the nations of Europe, to wit : by France, Belgium, Switzerland, Italy, the Pontifical States, Spain, Greece, Roumania, and Sweden, having an aggregate population of one hundred million, three hundred thousand inhabitants.

" The preliminaries of a monetary treaty, providing for a coin assimilating the ten florins of Austria with the twenty-five francs of France, were signed by the authorized representatives of those two nations on the 31st of July, 1867. It contained a condition in respect to the double standard, not yet complied with by France, but specimen coins were struck in anticipation, in October, 1867, bearing the heads respectively of the Emperors of France and of Austria, with the reverse, inscribed ' 10 florins, 25 francs.' The adoption by France of the single standard, to which a school of political economists in that country still remains violently opposed, but which has been repeatedly and earnestly recommended in the reports of official investigations (*enquêtes*) by large commissions of imposing authority, will secure the completion and execution of the treaty by Austria, and thereby increase the number of European nations using the franc to ten, with an aggregate population of one hundred and thirty five million, four hundred thousand inhabitants.

*   *   *   *   *   *   *   *   *   *   *   *

"The actual issue of the gold franc international coin has been commenced by some of the nations not included in the treaty of 1865. It is already in circulation in the flourishing principality of Roumania under the government of its present enlightened ruler, who has introduced in full the metrical weights, measures, and coins of France. His example, in all probability, will soon be followed by the Sultan of Turkey, who has manifested so strongly his wish to keep pace with the civilization of Europe.

"The assent of Spain to the franc system was given by the provisional government, in a formal decree, published at Madrid and Paris, soon after the expulsion of Queen Isabella.* A specimen of the new half franc silver coin now in possession of the undersigned, bears the well known 'pillars' with the ancient arms of Spain, encircled by the legend '400 *piezas en kilogramo*,' thus distinctly and arithmetically recording its assimilation with the 'franc,' two hundred of which are equiponderant with the French '*kilogramme*.'

"Among the northern European powers, the liberal and intelligent government of Sweden, so distinguished by its successful public works of intercommunication, has already issued the new gold pieces called '*Carolins*,' bearing the words '10 *francs*,' and now in common circulation in the countries on the Baltic. They are stamped with the 'image and superscription' of the Swedish monarch, CHARLES XV., lineally descended from BERNADOTTE and EUGÈNE BEAUHARNOIS, and now taking the lead in the civic march of monetary reform in the north of Europe. Under his influence and authority, Norway, which has a separate legistive body, will soon be united with Sweden in the gold franc system. His steady support of the legislative efforts of Mr. WALLENBERG, president of the Bank of Stockholm and one of the two delegates of Sweden and Norway in the Paris conference, has secured the coinage of a Swedish gold coin of twenty-five francs, to be issued as soon as similar pieces shall be put into circulation by the government of France."

These statements show a remarkable advance in public opinion since the fifth meeting of the International Statistical Congress, held at Berlin in 1863, at which time a measure so radical as the absolute unification of all currencies, seemed to the majority of that body altogether impracticable ; so that a committee of the congress, while proposing to reduce the units of money "to a small number," deliberately recommended that the pound sterling, the dollar, the florin, and the franc, should all

* These assurances were renewed to Mr. DE PARIEU, on a visit by him to Madrid in 1871.

continue to be retained. Even at that early day, Mr. Ruggles, who was present as the delegate to the Congress from the United States, protested against a conclusion so lame and impotent; and advocated universal unification on the basis of the gold unit of the value of five francs. He was thus the originator of the plan which, though defeated for the moment, was accepted later by the International Conference of 1867 ; and which has since been favored by the legislation of the continent to the extent above stated.

A proposition of this kind would, of course, not be made, unless the practical considerations in its favor were felt to outweigh any objections, derived from the beauty of congruity, or the exactions of a perfect theory. The arguments above cited therefore depend very much for their cogency upon the fact, that the gold coinage founded on the franc which is now actually in existence, constitutes nearly three-fifths of the total gold coinage of the world, and probably not less than two-thirds of that of Europe, including England. The unit of five francs, if adopted, will give at once an international character to this immense mass of coin, amounting to more than fourteen hundred millions of dollars—a fact of which the importance cannot be overestimated. It is, however, alleged in the report of the director of our mint, made in 1867, and also in statements by Baron Eugene Nothomb in the Prussian Annals, that the gold coinage of France is below standard weight (though within the limits of tolerance) even on its first issue from the mints ; and that it has even been designedly made so. The officers of the mint in Paris are said, however, to have denied this charge, and to have claimed that their coinage not only falls very little below the standard, but often exceeds it. Opinions must therefore be suspended ; but if the statement be well founded, it must tend to diminish the weight of the argument derived from the magnitude of this coinage. To the people of Great Britain and of the United States, whose coins, if this unit were adopted, would require to be recoined, the latter being more than three and a half per cent. in excess of weight, the plan will only become acceptable, when it shall havo been clearly demonstrated to them that no other possibility exists of reconciling the monetary systems of the world.

The second of the expedients enumerated above as proposed for securing an international currency, viz., that of adopting tho unit of weight, or one of its decimal derivatives, as the visible representative in gold of the unit of money, was suggested by Mr. MICHEL CHEVALIER, in the first instance, in the *Journal des Economistes*, for November, 1868, and afterwards advocated by him in the French Senate, as appears in the debate above cited. He proposed to take as a unit the *gramme* of standard gold, nine tenths fine, of which the value in our currency would differ but by an inappreciable fraction from sixty cents; or the dekagramme, which would be equivalent, with a correspondingly slight difference, to six dollars. Calling the coin of one gramme weight the *soldo*, the dollar would become, by an insignificant modification, one soldo and two thirds of a soldo ; or six dollars would be equal to ten soldos. Our gold coins might therefore continue to serve as money without inconvenience, until such time as, having been worn out, they should be replaced by other coins in the form of soldos and multiples of the soldo. The napoleon, increased in weight by less than three fourths of one per cent., would be equal to six and a half soldos. The pound sterling, diminished in weight by one and two thirds per cent. would have the value of eight soldos. Thus the coinage of the three principal commercial nations of the world could still be made to subserve the purposes of money, till such time as a new coinage should be created to take its place. But these three nations, and all others adopting the system, would be obliged at once to change their unit of account, and with it their whole monetary system.

This scheme certainly has its advantages. It is strongly advocated by Dr. FARR, the ablo delegate from Great Britain to tho

International Statistical Congress held at the Hague in 1869 ; though with a preference for the dekagramme, of the value of six dollars, or of one and a quarter (modified) sovereigns, as a unit, which he would name the *Victoria*. Five sovereigns on this plan would be equal to four Victorias.

The third of the methods suggested as offering a practicable solution of the problem under consideration, is exemplified in a proposition made in 1868 by Mr. GEORGE F. DUNNING, Superintendent at that time of the United States Assay Office in New York. This is to assume a unit having a weight of one gramme and sixty-two one-hundredths of a gramme of gold, nine-tenths fine ; which corresponds, with a fractional difference entirely inappreciable, to twenty-five grains Troy. As a dollar weighs twenty-five grains and eight-tenths, this unit is but about three and one-tenth per cent. less, and it is proposed to call it still a dollar. The piece of five francs increased by forty-four one-hundredths of one per cent. becomes then equal to a dollar, and the sovereign diminished in the same proportion becomes the equivalent of the half eagle. This scheme has simple relations to the grain weight; but this recommendation is likely to lose its force with the prospective probable abandonment of the use of that weight altogether. The change which it proposes in the weight of the French coin, though not large, is in a direction unfavorable to the preservation of that coin long in use, if, as affirmed, the coinage is already too light ; and it diminishes materially the value of our dollar without any corresponding advantage, which would not be gained by making it equal at once to the piece of five francs.

Another proposition founded upon the same principle, but recommended by a simpler relation to the metric system of weights, and a close approach in the unit proposed to the value of the American dollar, was presented in a petition to Congress from the American Statistical Association in 1867, and was laid before the American Association for the Advancement of Science at its meeting of 1868 in Chicago, by Mr. E. B. Elliott, of the United States Treasury, with whom it originated. This consists in making the dollar of such weight of standard gold as to contain one gramme and a half of the pure metal. The proposed alloy being one part in ten, it follows that the total weight of the dollar will be one gramme and two thirds; or employing the word *tergramme* to signify one third of a gramme, the weight of the dollar will be five tergrammes.

The division by three is justified, and even recommended, by the consideration that the presence of an alloy in the metal of coinage precludes the possibility of a strict adherence to the decimal scale. As it is probable that the fineness of nine-tenths will ultimately be universally preferred to any other, it will follow that ten parts of the alloyed metal will contain nine of the pure metal; which latter is only accounted to have value. And the simple divisor of nine is three.

This unit is less in value than the American dollar by a little more than three-tenths of one per cent.; a difference which is below the limit fixed by law for the tolerance of variations in weight, in the half eagle and inferior coins; so that the existing coinage of the United States would not be displaced by it. Its adoption in England would require an increase of weight of two .and forty-three one-hundredths per cent. in the sovereign; and its introduction into France would necessitate a change in the napoleon in the same direction, to the extent of three and thirty-three one-hundredths per cent. Affecting, therefore, so seriously a very large proportion of the gold coinage of the world, this unit is likely to make its way but slowly, if at all. It is, nevertheless a more convenient unit of account than the pound sterling or the franc; it is simply related to the metric system of weights, which bids fair to become in due time universal; it is very nearly the equivalent of the Spanish dollar, so extensively employed in the commerce of the far east; it is an exact submultiple of the triple crown of the German Muntzverein;*and it

* This coin has practically disappeared from circulation.

is not so largely discordant with the coinage systems of England and France even, as to exclude the coins of those countries from circulation ; though, till replaced by others, such coins might circulate at a discount.

In the notice which has here been taken of the several plans, which up to this time have been seriously advocated, for the unification of the monetary systems of the world, it has not been intended to indicate a preference for one of these rather than for another ; nor even to assume that the possibilities in the case have been exhausted, or that a better plan than any one of them may not yet be proposed.   The object in view has been simply to point out the peculiar difficulties which surround the subject, and to illustrate their magnitude ; so that the propriety of dealing separately with the question of weights and measures may not seem doubtful.

To the plan of unification which rests on the French gold piece of five francs as its basis, the nations parties to the monetary treaty of December 23, 1865, including France, Belgium, Switzerland, and Italy are already prepared to assent.   These nations have, in fact, only to abolish the double standard, and they have the system already   The treaty referred to establishes for the parties to it a common system of coinage, to be maintained until January 1, 1880, if not sooner repealed.   Spain, Sweden, Austria and Greece have exhibited a willingness to accept the same system ; and Roumania has put it into practice.

The system of which the basis is the five tergramme coin of gold, of the fineness of nine-tenths, exists, as we have seen, practically, in the United States, at the present time.   An important and rather unexpected encouragement has been afforded to the advocates of this scheme, during the current year, by the action of the government of the Japanese Empire.   Sometime during the year 1870, the Master of the Mint of that Empire was deputed as a commissioner to visit Europe and the United States, with instructions to inquire into the systems of currency and

coinage prevailing among different peoples, in the design to effect
a complete remodelling of the coinage of Japan. Commissioner
Ito, after having fully informed himself of the diversity of views
entertained by the prominent authorities on the subject in the
countries visited by him, returned at length prepared to recommend, first, an entire departure from the traditions of the East
as to a standard metal, by discarding the silver standard, and
adopting the standard of gold only; and secondly, the adoption,
as the unit basis of the new system, of a gold coin of the weight
of one gramme and two-thirds, of the fineness of nine-tenths.
The views of the Commissioner were approved ; and a decree of
the empire, which was promptly issued in conformity with them,
provides for the stamping of gold coins of the several values, one
yen, two yens, five yens, ten yens, and twenty yens—the yen being
the name given to the unit-base or gold dollar. This coinage is
made the only legal tender in the empire ; except for small payments, which may be made in fractional coins of a subsidiary
silver currency ; and except also for transactions in the open
ports, for which a silver one-yen, the equivalent of the Spanish
dollar, continues to be coined, and is made legal tender for all
sums.

This remarkable action, which illustrates in a striking manner
the energy, the independence, and the freedom from prejudice,
with which the sagacious rulers of the Japanese islands are seeking to improve the social, political, and material condition of
their people, is possibly destined to have a larger influence on the
ultimate settlement of the question of an international coinage
than immediately appears. Should the example of Japan be
followed by the neighboring and more populous empire of China,
the demand of Oriental commerce may hereafter be for gold dollars instead of for silver; and this demand may decide the question what precise weight can be most advantageously given to the
gold coins of Europe and of America.

The latest incident of importance which has occurred in the

political world, affecting the question of monetary unification, is one which rather increases than diminishes the difficulties surrounding this question; and one which we are therefore compelled to regard as a serious misfortune. During the closing months of the year 1871, Germany has permitted to pass unimproved, or rather has put to a very bad use, an opportunity such as can rarely be presented to a people for securing a great public benefit—an opportunity which she might have improved in such a manner as to settle decisively, at least for the continent of Europe, the question of an international coinage. While these lines are in writing, intelligence is received of the adoption of a monetary gold unit for the empire, upon a new basis called the *mark*, supposed to be equivalent in present value to one third of a Prussian silver *thaler*. The principal coin of this system is to contain twenty marks of gold, nine-tenths fine, and to have the weight of seven thousand, nine hundred and sixty-five milligrammes :—with the value, in the currency of the United States, of 4.76426 dollars.* This coin not being commensurable in weight or in value with either the franc, the dollar, or the sovereign, serves only to embarrass still further a problem already sufficiently difficult ; and its introduction must have the effect to postpone indefinitely, for central Europe at least, the enjoyment of the great blessing of an uniform international currency, of which the prospect seemed recently so promising.

The heavy discouragement which this untoward action has brought with it to the cause of uniformity is, however, measurably relieved by some attendant and partially compensating advantages. If Germany has failed to grasp the favorable occasion for reconciling the conflicting monetary systems of Europe,

---

* The law provides for the minting of two gold coins ; a piece of ten marks, and a piece of twenty marks. One hundred and thirty-nine and a half ten-mark pieces, or sixty-nine and three-quarters twenty-mark pieces are to be made from one German pound (five hundred grammes) of fine gold. A mark has the value of 23.8213 cents of the currency of the United States.

she has nevertheless brought order out of the confusion of her own. Henceforth, in reference to this matter, she contributes but a single element instead of many to the general discord; and permits us to say that the number of the inharmonious elements which, in studying the question of unification, we are compelled to consider, is practically reduced to these four—the mark, the franc, the pound sterling and the dollar. The relations which these bear to each other, as presented in the weights and values of their characteristic gold coins most nearly approximate, are given here—the weights being expressed in grammes and milligrammes, and the values in the currency of the United States :

Country.	Coin.	Weight. Pure gold.	Weight. Standard gold.	Value.
Germany, 20-mark piece.........		7.168	7.965	$4.76.4
France, 25-franc piece (proposed)		7.258	8.065	4.82.4
Great Britain, Sovereign.........		7.322	8.136*	4.86.5
United States, Half Eagle........		7.523	8.359	5.00.0

In this exhibit we have the whole case immediately under the eye. Till Germany intervened, the widest range of difference between the representative coins of the systems which it seemed most important to reconcile, was less than eighteen cents. She has extended it to nearly twenty-four. If the hope of an early and general unification seemed doubtful before, it must now be pronounced little better than desperate.

Yet it by no means becomes the friends of monetary reform on this account to fold their arms in discouragement. The very hopelessness of the case in the aspect here presented, imposes upon the four great powers above named, the moral obligation to lend their most serious attention to every alternative possibility of relief which may be suggested ; and, in the actual state of things it is eminently well worth their consideration, whether present effort should not be directed

---

* Reduced to fineness of nine-tenths; present weight, eleven-twelfths fine, 7.988 grammes.

toward the creation of a coinage for international use upon a plan entirely independent of all local and purely national systems. Such an international coinage, though not in strictly commensurable relations with the coins now actually in use, would be productive of the great benefit of immensely simplifying commercial exchanges, and of providing travellers with a currency everywhere invariable in value. The advantages which its introduction would bring with it, are to some extent even now actually enjoyed throughout a large part of Europe, in consequence of the extended circulation which the gold napoleon has secured beyond the limits of the French territory. This coin is in fact international with the nations parties to the monetary treaty of December, 1865; and it is to this fact, no doubt, that it owes much of its popularity elsewhere. But though it circulates freely all over Germany, Austria, Holland, Denmark, and Sweden, it circulates only by toleration, and because of its great convenience ; and is not a legal tender for payments of any amount, in any of those countries.

The international coinage here suggested will command, if established, an immediate circulation among all the peoples who shall unite in its creation ; and will bring with it greater advantages than have attended the continental circulation of the napoleon. The existence of such a coinage will moreover keep constantly before the minds of all those who use it, the desirability of a currency which shall be not only international but everywhere uniform and identical. Thus the proposed coinage may become a powerful instrumentality in promoting the attainment of the object for which it seems at first to be only designed as a substitute—the ultimate complete unification of the monetary systems of the world.

An international coinage devised without reference to systems actually existing, and controlled in its principles by no considerations of what may be their discordances or resemblances, may be moreover constructed on a strictly scientific plan; and thus,

when time shall have familiarized the nations with its values
and its visible forms, it may happen that, without violence,
and by the spontaneous action of peoples themselves, diversity may silently give place to uniformity, and a better system than any which present legislation could create, may
ultimately prevail throughout the world.

NOTES SUPPLEMENTARY.

1.—*Effect upon existing Contracts of a Change in the Legal Weight of Coins.*

Since committing the foregoing pages to the press, it is felt that sufficient prominence has
not been given to the thought, cursorily suggested on page one hundred and thirty-one, that
every change, small or great, introduced into the weight of legal tender coins, while the same
coins continue to be unchanged in nominal value, imposes upon the government making it, the
most stringent obligation, in honor and in morals, to provide for the full and exact fulfilment
of all outstanding pecuniary engagements in the sense in which they were understood at the
time they were contracted. From this obligation there is no possibility of escape. The public
opinion of the world, to which in this age even monarchs must defer, will not suffer that, in
the light of the nineteenth century, governments shall presume, on any pretext whatever, to
imitate practices such as those to which unscrupulous despots were accustomed to resort
in ruder times, that they might wring from the hand of honest industry its hard earned
wages. There is one thing that is more valuable than national wealth or national power,
and that is national honor; and it would be better to endure, to the end of time, all the evils
and disadvantages with which the diversity and incongruity of existing monetary systems
embarrass intercourse and encumber commerce, or even to go back to the rude iron currency
of the Spartans, than to possess ourselves of the most perfect system that science can devise,
through the operation of a legalized injustice. There can be little doubt that it was the
gravity of these considerations which determined the government of the United States to
make to the governments of Europe the proposition mentioned on the page above referred to; the effect of which would be, in case of its acceptance by those governments,
to acquire to some extent for existing coinages the character of an international coinage, by making the values of their principal coins commensurable.

This plan, if adopted, owing to the minuteness of the changes which it demands, would
simplify the legislation required for the introduction of an international coinage, and would
secure the people, whom such legislation affects, against the inconvenience arising from the
necessity of discharging their pecuniary liabilities in a currency differing from that which
existed at the time when those liabilities were incurred.

2.—*The New System of Coinage of the Japanese Empire.*

In addition to the statements made on page 147, the following particulars may interest. The fineness of the subsidiary silver coinage is eight-tenths ; and the ratio in value of the silver in this to gold is 1 : 13½. This coinage is legal tender to the amount of ten yen (dollars). In the silver one-yen coin for the open ports, the fineness is nine-tenths ; and the ratio 1 : 16.173. (The gold one-yen contains 1.5 grammes of pure gold, and the silver one-yen, 24.260726 grammes of pure silver.)

The gold coins are as follows : one yen, two yen, five yen, ten yen, and twenty yen. These are all legal tender for all payments, to any amount.

The rules relating to the silver one-yen are peculiar, and are as follows :

" For the sake of rendering facility to foreign commerce at those ports open to foreigners, and in accordance with the requirements of the traders, both foreign and Japanese, our government will coin the silver one-yen and make it useful only for foreign commerce. This silver one-yen will be the legal tender in payment of local taxes and of import and export duties. It will also be the legal tender in any commercial transaction in the open ports. The said silver one-yen will, however, not be the legal tender in any other place than the said open ports, and shall not be used for the payment of internal revenue of any kind, nor will it be lawful currency in the interior : but by mutual agreement any person or persons may use it to any amount throughout Japan.

" In the payment of the import and export duties, the comparative rate of the gold yen to the silver one-yen will be as follows : thus, one hundred silver one-yen shall for the present be equivalent to one hundred and one gold yen."

The world will wait with interest for the result of this curious experiment. It need hardly have been provided that the silver one-yen shall not be legal tender in the interior. The probabilities are that it will never be seen there ; and it is furthermore probable that, so long as the legal relation between the silver one-yen and the gold yen remains as provided in the last paragraph foregoing, all the silver one-yen coins that can be produced will be promptly bought up and exported as fast as they are issued from the mint.

APPENDIX B.

Note 1.—ON MEASURES OF CAPACITY, AND THE WEIGHT OF A GIVEN VOLUME OF WATER.

(REFERRED TO ON PAGE 39.)

A measure of capacity is, theoretically, a certain defined volume expressible in cubic units, integral or fractional, derived from some invariable linear measure. In theory, therefore, the measure of capacity is the same under all conditions. But, in practice, the capacity measure, being a vessel of metal, wood, or other material subject to expansion and contraction with variations of temperature, can correspond strictly with the theoretic measure only at some particular temperature.

In the United States the standard temperature is that at which the density of pure water is maximum. This is stated by Mr. F. R. Hassler, by whom, as Chief of the Bureau of Weights and Measures, the experimental investigations were conducted which resulted in fixing the methods of adjusting the measures of capacity, in a report made by him, Jan. 27, 1832, as being true at 39°.83 * of the thermometer of Fahrenheit. The capacity of the bushel was, at the same time, fixed at 2150.42 cubic inches, which is the capacity of the old Winchester bushel of England; and that of the gallon, at 231 cubic inches, corresponding to the British wine gallon. But, inasmuch as the British standard temperature is 62° F., the Winchester bushel

---

* In his report of 1842, ten years later, Mr. Hassler tacitly admits that this temperature is too high, by speaking of 39° F., as the temperature of maximum density. Playfair and Joule make it 3°.945 C. = 39°.101 F.

of England and the Winchester bushel of the United States are equal to each other neither at 39°.83 F., nor at 62° F., nor at any other common temperature. In point of fact, the British Winchester bushel, when made of brass, contains, according to Mr. HASSLER, only 2148.9 cubic inches at the temperature adopted as standard in the United States ; falling short of ours by one cubic inch and fifty-two one hundredths ; while ours exceeds the British by the same amount, at the temperature according to which the standards of Great Britain are adjusted.

To construct a measure of any description, whether of length, or of capacity, or of weight, in exact conformity with a standard previously imposed, is a practical problem of extreme delicacy and difficulty. The verification of capacity measures is moreover attended with the special disadvantage, that the interiors of vessels cannot be subjected to microscopic measurement, with the same facility as the external contours of solids. Capacity measures designed to serve as standards are therefore verified by an indirect process, which consists in determining, by means of the balance, what quantity of pure water of known temperature they are capable of containing.* But in order that this method may be worthy of reliance, it is necessary first that the absolute weight of a given volume of water should be ascertained with the utmost exactness ; and this not only at a given temperature, but at every temperature within a pretty wide range ; since the rate of expansion of water with increasing temperature from the point of its maximum density, is governed by no obvious law. To the determination, therefore, of the

---

* During the process of weighing, the measure is closed by means of a plate of truly plane ground glass, the top or rim being correspondingly ground. It would not, of course, be possible, by any exercise of dexterity, to fill an open vessel exactly full of liquid, and no more ; nor, supposing that condition reached, could it be verified by simple observation. Should any air bubbles be inclosed within the covering, they are brought by management to the centre of the plate, where there is a small perforation, through which the deficiency of fluid may be supplied. The filling having been perfectly accomplished, the exterior of the vessel and cover are then carefully dried before weighing.

weight of a cubic inch, or of some other determinate volume, of water, there has been devoted a vast amount of laborious experimental inquiry ; with results which differ only by a few one-thousandths of a grain per cubic inch, and which are at length probably as near to accuracy as it is possible for human skill to attain.

In the prosecution of this inquiry the method pursued has not been to prepare first a vessel of measured capacity, and then to ascertain by the balance what weight of water in ounces or grains it will hold. The difficulties already mentioned in the way of constructing capacity measures by actual direct measurement, prevent this. But the plan has been to prepare a solid of regular geometrical figure, and of specific gravity superior to water ; and then to find experimentally the apparent loss of weight which this body undergoes, when immersed in water. The dimensions of such a solid can be determined with great accuracy. It displaces precisely its own bulk of the fluid ; and the observed loss of weight is therefore directly the thing sought —that is, the weight of the same bulk of water.

The weight of the cubic inch of water at maximum density, as accepted at the Bureau of Weights and Measures of the United States, is stated by Mr. HASSLER, in a report to the Secretary of the Treasury, dated March, 1842, to be 252.7453 grains; and this is there said to have been deduced from the " English determinations ;" and to be the value according to which the measures of capacity of the United States are adjusted. It is a little curious to observe, however, that the same authority gave, in his report of 1832—the report in which the regulating weights of those measures of capacity were first declared and prescribed —a materially different value of the same constant. In that report he states the capacity of the gallon to be two hundred and thirty-one cubic inches ; and the weight of two hundred and thirty-one cubic inches of water at maximum density to be 58,372.1754 grains ; from which it appears that the cubic inch of water at maximum density was supposed by him then to weigh only 252.6934 grains, as given in the text of the foregoing address. This, however, is not a British determination; but it is Mr. Hassler's own, deduced from the weight of an authentic kilogramme, described by him to be "an original kilogramme of brass, standarded by the Committee of Weights and Measures in Paris [the committee on the 'definitive metre'] exactly equal with all those of the deputies of foreign nations present at the committee ; " combined also with "an authentic metre of the committee," which he had himself twice compared.   This value of the kilogramme is 15,433.1669 grains, exceeding that found by Prof. Miller, in 1844 (the value given implicitly on page 15), by 0.81816 grains.   Mr. Hassler's determination of the metre was 39.3810327 inches, exceeding that received at the British Standard's Office, as established by Kater in 1818, by the small fraction of an inch represented by the decimal 0.0102427.   If we employ the numbers of Miller and Kater, we shall find the weight of the cubic inch of water *in vacuo*, at maximum density, as deduced from the kilogramme, to be 252.87718 grains.   Supposing the weighing to take place in air, it will be necessary to deduct from this, the weight of the air which the water displaces, diminished by the weight of the air displaced by the counterpoises. In the British computations, the specific gravity of the counterpoise weights is taken at 8.5.   The allowance for buoyancy must, according to this, be taken at $\frac{7.5}{8.5}$ ths of the weight of a cubic inch of air ; or of a cubic inch of the water weighed, multiplied by the specific gravity of the air at the time of the experiment, taken with reference to water of that density.   According to Mr. Hassler, the value of this specific gravity which was employed by the French committee was $\frac{1}{770}$; and if we compute the correction upon this specific gravity, we shall find it to amount to 0.28978; so that the resulting weight of the cubic inch of water at maximum density, weighed in air at 39°.83 F. and at 30 inches of barometric pressure, will be 252.5874.

This specific gravity is, however, too large, as is shown more at length in Note 2, following. It should be put at 0.001269582 $= \frac{1}{787.66}$; and if this value be employed, the correction will become 0.28345 ; and the weight of the cubic inch of water in the air at normal pressure, deduced from Miller's weight of the kilogramme, will be 252.59373 grains.

Hassler's weight of the cubic inch, as derived from his own authentic kilogramme, must be regarded as a weight *in vacuo ;* although he treats it as a weight in air, and makes it the basis of his methods of adjusting the standards of capacity of the United States ; in which adjustments the weighings take place, of course, in the air. He has not informed us whether, when his brass kilogramme was compared with the platinum standard, allowance was made for displacement ; but we should naturally suppose that that was done, and the agreement of the weight with the weight of the prototype of the Archives, as found by Prof. Miller, within about eight-tenths of a grain, is in favor of the supposition. Moreover, the effect of displacement in this comparison could only be slight. According to the determinations given in Note 2, following, the weight of air displaced by the brass kilogramme at the temperature of 62° F., and under 30 inches of pressure, would be only 2.20693 grains, and that of the platinum kilogramme 0.89328 grains. The resultant effect would therefore be 1.31365 grains ; an amount by which Mr. Hassler's kilogramme ought to be increased in order to give the weight *in vacuo*, if the allowance was not originally made.*

But that the allowance *was* made can hardly be doubted; since Mr. Hassler says, not that his brass kilogramme was *compared* merely with the platinum prototype, but that it was " standarded" by it—and that this was done, moreover, by the original standards-committee of 1799. The weight of the cubic inch of

---

* In determining the weight *in grains* of the brass kilogramme, after the comparison with the prototype, no allowance for displacement was necessary, the counterpoise weights being of the same metal.

water found from it, by dividing by the number of cubic inches in the cubic decimetre, as derived from his own authentic metre, which weight is 252.6934, is therefore a weight supposed to be taken *in vacuo* at the maximum density. This weight will be sensibly increased, if the cubic decimetre as determined by KATER (now generally received) be made the divisor ; becoming in this case 252.89065 grains.

That Mr. HASSLER should have taken this weight, which is thus seen to be a weight *in vacuo*, as the basis of the adjustments of the national standards of capacity by weighings in the air, is a circumstance too remarkable to be credited, except upon the most conclusive evidence. His report of 1832 leaves, however, no doubt upon that point. In that report, after having stated that water at its maximum density " is very properly chosen to determine the bushel and gallon by the weight of distilled water which their legal capacity will contain," he proceeds to add : " Thus were determined, in weight of the Mint in Philadelphia, to contain distilled water at the maximum density, and at thirty inches barometer, the bushel =543,391.89 grains = 77.627413 pounds avoirdupois ; the gallon =58,372.-1754 grains = 8.33888220 pounds."

If we now divide 58,372.1754 by 231, the "legal capacity" of the gallon in cubic inches, the quotient is exactly 252.6934, without remainder. If we divide 543,391.89 by 2150.42, the capacity of the bushel, the harmony of the two statements is not at first apparent ; the dividend wanting a little more than five grains in order to furnish the same quotient ; but we presently perceive that Mr. HASSLER has given to the bushel only 2150.4 cubic inches, dropping the last decimal ; and that 252.6934 $\times$ 2150.4 = 543,391.88736 ; or, taking the nearest deci-mal in the hundredths place, 543,391.89, as stated in the report.

The value given, as mentioned above, by Mr. HASSLER, in his report of 1842, as being " the English determination," and which is there stated by him to be the basis on which the United States measures of capacity are founded, is also given explicitly in the report of ten years before—*i. e.*, the same report in which, as we have seen, the standards of capacity are regulated in accordance with the weight derived from the kilogramme ; so that the choice at that time seems to have been deliberately made. The difference between the values is not of serious importance, amounting only to eleven or twelve grains in a gallon ; but it is certainly a logical error to found a measure of capacity which is to be adjusted by weighing in the air, upon a unit of weight which is true only *in vacuo.* It was perhaps also an error of judgment to derive this unit of weight from the kilogramme at all ; when the remaining determinations cited in the report in which this choice is made, are discordant with this, and are generally harmonious among themselves.

It might be supposed that when Mr. HASSLER, in 1842, announced that the capacity measures of the United States were adjusted by weight in accordance with the English determination of the weight of the standard cubic inch of water, viz., 252.7453, he had modified the total weight of the gallon correspondingly; and this, if he had been silent as to that point, would necessarily have been taken for granted. He removes, however, the possibility of such an impression, by immediately stating, in the same report of 1842, that the weight corresponding to the gallon is 58,373 grains ; a weight identical with the original determination, excepting that the decimal is dropped, and the last figure of the integral number is increased by a unit. If the British value had been actually employed, the gallon, in order to preserve its "legal capacity," should have been made to contain 58,384.1643 grains ; or 11.9889 grains more than assigned it in 1832.*

---

* On supposition that the value 252.7453 grains is true, the actual standard brass gallon measure of the United States has the legal capacity of 231 cubic inches, not at the temperature of the maximum density of water, but 6°.51 F. above that temperature ; or, putting the temperature of maximum density at 39°.1 F., the brass standard gallon contains 231 cubic inches at 45°.61 F.

The repôrt of 1832 gives not only the determination called the English; but also the manner in which this value was deduced. It is a mean of several, all of which were obtained by reducing the weight given in the British act of Parliament of 1824 fixing the standards of capacity, as being that of a cubic inch of water taken at 62° F. of temperature and under a barometric pressure of thirty inches, to the temperature of the maximum density of water ; by means of coefficients, experimentally determined, for the variation of density with changes of temperature. The value given in the act is 252.458 grains. It was derived originally from the very elaborate researches upon weights and measures of Sir GEORGE SHUCKBURGH EVELYN, a member of the Royal Society ; of which a full account was published, in 1798, in the *Philosophical Transactions.* In determining, in the course of these researches, the weight of a given volume of water, Sir GEORGE made use of three different solids; one of them being a cube, one a cylinder, and one a sphere. In the year 1818, a Royal Commission was appointed and charged with the duty of inquiring and reporting on the subject of weights and measures, with a view to legislation. The chairman of this committee was Sir JOSEPH BANKS ; and it embraced, among other distinguished members, Dr. WOLLASTON, Dr. YOUNG, and Captain HENRY KATER. The experimental inquiries instituted by the committee were chiefly conducted by Capt. KATER; and, in the course of them, this gentleman carefully revised the researches of Sir GEORGE SHUCKBURGH, remeasuring with microscopic accuracy the three solids which had been used by him. The weighings were not repeated. They had been made by means of a set of weights constructed expressly for the purpose by TROUGHTON; which weights accorded perfectly among themselves, but were a trifle light when compared with the parliamentary standard; the difference amounting to nearly two thirds of a grain in one thousand grains. The weighings were made in the air at different temperatures and under different pressures, the temperature of the water differing also from that of the air; but the results were all reduced to the standard temperature of 62° F., and the standard pressure of thirty inches of the barometer. The volume of the cube was 124.1969 cubic inches, at 62° F. of temperature; that of the cylinder, 75.2398 cubic inches; and that of the sphere 113.5264 cubic inches. The first gave the weight of the cubic inch of water, *in vacuo*, 252.907 of Sir GEORGE SHUCKBURGH's grains; the second 252.851, and the third, 252.907. The mean of these values is 252.888; which reduced to grains of the parliamentary standard is 252.722.

In making these reductions, every circumstance which could affect the result was carefully considered ; the difference of bulk of the solid at the temperature of the experiment and at 62° F., as well as the difference of density of water at the same two temperatures, being computed with severe accuracy. The buoyant power of the atmosphere in diminishing the weight, not only of the solid immersed, but also of the counterpoise weights, was likewise taken into the account.*

The weight in the air deduced from the weight *in vacuo* given above, is obtained by subtracting the calculated weight of the air displaced by the water, less that of the air displaced by the counterpoises. The remainder, being multiplied by a coëfficient expressive of relative density at the two temperatures compared, gives the weight at the temperature of maximum density, and under the standard pressure.† Mr. HASSLER's report of 1832 gives three such coëfficients, derived directly or indirectly from GILPIN, with the weights obtained by means of them ; and also the mean of these weights, which is 252.7487 : but the particular result which is called, in the subsequent report, "the English determination," viz., 252.7453, is one given by VAN DER TOORN, of Amsterdam, from his own computations, based on GILPIN.

---

* From an examination of KATER's computations, it appears that the constants employed by him may, in some instances, be advantageously modified. His reductions have been, therefore, carefully re-examined, and the results are given in a separate note, which follows the present.

† This was Mr. HASSLER's order of proceeding : rigorous accuracy would require, however, that the reduction to maximum density should be made before correcting for buoyancy.

In the same report are also given four results from French
authorities, with a mean of 252.7098 ; and quite a variety from
VAN DER TOORN, ranging from 252.5658 to the one given above,
viz., 252.7453. The results from British authorities, of which
the mean is given in the last paragraph, are all higher than any
from other sources. Of fifteen different determinations embraced
in the table of the report, all except five are greater than that
upon which Mr. HASSLER founded the capacity measures of the
United States.

It is difficult to account for the not inconsiderable difference
which appears, from the foregoing figures, to exist between the
weight of the actual prototype platinum kilogramme of France,
and the weight of the theoretic kilogramme, as deduced from the
British experiments. The original determination of this standard
was made with great care, in 1795, by M. LEFÈVRE GINEAU ; the
method employed by him having been the same which has been
described as used by Sir GEORGE SHUCKBURGH. The solid employed
in the weighings was constructed in the form of a cylinder, such
a body being capable of being constructed and measured with
the greatest precision. It was hollow, being made only heavy
enough to sink freely in water. In dimensions it was nearly
two decimetres and a half in diameter, and of equal height. The
exact mean height and diameter were determined by an extreme
refinement of measurement to the ten thousandth of a millimetre, at the temperature of 17°.6 C = 63°.68 F. The volume
computed upon these measurements was then reduced to the
temperature of melting ice, at which it was determined to be
very nearly 11.28 cubic decimetres = 688.385 cubic inches.

The weighings were made by means of weights prepared expressly for the purpose ; the unit being an approximate kilogramme, and the smaller weights, decimal sub-multiples of this
down to the one-millionth part, carefully verified, and pre-
sumed to be less than one half of one one-millionth part in error.
The counterpoise weights having been of the same metal with the

cylinder, no allowance for displacement was necessary in weighing in the air, the interior of the cylinder being in communication with the atmosphere during this operation. In weighing the cylinder immersed, the weight of the confined air was deducted as well as that of the air displaced by the weights. Thus, no precaution was neglected which seemed to be necessary to secure accuracy; and yet the resulting weight of the cubic decimetre of water as thus determined is sensibly less than that deduced by the British and by most of the continental experimenters.

Mr. CHISHOLM gives, however, in the Sixth Appendix to the Second Report of the British Standards commission, " an important computation of the actual weight of a cubic decimetre of water at its maximum density, in terms of the standard *Kilogramme des Archives*, the computation being based upon KUPFFER's observations of the weight of a cubic inch of water and of the standard kilogramme." The details of these observations and calculations are given in a voluminous report of the commission appointed to fix the standards of weight and measure of the Russian empire, published in 1841.

KUPFFER appears to have found the weight of the cubic inch of water *in vacuo*, at $13°\frac{1}{3}$ R. $= 16°.67$ C $= 62°$ F., to be only 252.598 grains. He also found the weight of the standard kilogramme to be 15432.36186 grains ; exceeding that found by Prof. MILLER by the minute fraction 0.01312. Mr. CHISHOLM remarks that, " according to this computation, based upon KATER's valuation of the metre, a cubic decimetre of water at its maximum density weighs 1000.0115 grammes ; and therefore 11.5 milligrammes, or 0.17747 grains more than a kilogramme. By a similar computation, if Capt. CLARKE's more recent valuation of the metre $= 39.370432$ English inches be taken as the base, a cubic decimetre of water, at its maximum density, weighs 15 milligrammes, or 0.23145 grains *less* than a kilogramme. If the mean of these two computations be taken, the weight of a cubic decimetre of water at its maximum density will be only 1.75 milligrammes, or 0.027 grains less than a kilogramme."

Mr. CHISHOLM has carefully verified all the computations of KUPFFER, and remarks that "KUPFFER's operations for determining the weight of a cubic inch of water appear to have been made with the greatest possible care and accuracy." He concludes, however, with the observation, that "in point of fact, the authority of a computation based upon the imperial weights and measures used by KUPFFER cannot be put in comparison with that based on the imperial *standards* used by SHUCKBURGH and KATER;" and that "it may therefore be stated that the legal weight in a vacuum of a cubic inch of water at 62° Fahrenheit, thus scientifically and authoritatively determined, is 252.722 grains."

One further remark remains to be added, in order that the purpose of this note may be complete; which is in reference to the mode of adjustment of the standard measures of capacity actually employed in practice at the Bureau of Weights and Measures. Though the weight of water which these standards should contain is that of a definite number of cubic inches taken at the temperature of maximum density, and under the barometric pressure of thirty inches, yet the adjustments are made at the ordinary temperature, and in any state of the barometer; the test weight which the vessels are made to contain being that which the expansion of the material from the standard temperature to the temperature of observation would adapt them to hold of the fluid, which also expands simultaneously. If it were true that the capacity of the vessel and the bulk of the contained liquid expanded equally throughout the range to which observation extends, then it would happen that the vessel would hold exactly the same weight of liquid under all variations of temperature. But this is not the case. Near the standard temperature, the capacity is enlarged by elevation of temperature more rapidly than the bulk of the liquid; and, therefore, for a range of about twelve or thirteen degrees above the temperature of maximum density, the vessel is capable of holding a somewhat larger weight than at that temperature. But above the forty-sixth degree of Fahrenheit's thermometer, the volume of the liquid begins to expand faster than the capacity of the vessel ; so that, at about 52° F., the weight of the water which the vessel will contain is once more equal to what it was at the standard temperature. This makes the temperature of 52° F. a particularly favorable one for the practical operations of the adjustment ; and it is one which is more easy to command than that of 39° F.

The barometric corrections, according to Mr. HASSLER, are made on the supposition of a constant specific gravity of $\frac{1}{831}$, of air, under the normal pressure. In NOTE 2, which follows this, are stated some reasons for believing that this specific gravity is too small. By the method there described, the weight of a cubic inch of air of ordinary humidity under the pressure of 30 inches, at 52° F., is found to be 0.31124313 ; and this, divided by the weight of a cubic inch of water *in vacuo* at 62° F., as there found, viz., 252.75965, gives the specific gravity 0.001231400 $= \frac{1}{812.083}$. As in practice, however, the specific gravity appears to be made a factor in a coëfficient of correction, which is multiplied directly into the observed weight of the water in the vessel, the divisor should be the weight *in vacuo* of the cubic inch of water at 52° F. rather than at 62° F. ; which weight is 252.94021 grains, giving a specific gravity of 0.001230501 $= \frac{1}{812.677}$. But as it is, probably, the water only which is taken at 52° F.; the weighing being made in air of ordinary temperature, or between 60° F. and 70° F., Mr. HASSLER's specific gravity will, under the actual circumstances, be very near the truth. It is however to be observed that, if the normal weight of the cubic inch of water at maximum density in air, be taken on supposition that the air is itself also at the temperature of 39.°1 F., there should be a correction for the thermic condition of the atmosphere, as well as for pressure : and as this would, in general, be materially the greater of the two, it is apparently a needless refinement of accuracy to consider the fluctuations of the barometer, if the atmospheric temperature be disregarded. A cubic inch of air, under normal pressure, weighs, at 62° F., about one twenty-third part less than at 39° F. ; while even so large a barometric variation as an entire inch, affects, at constant temperature, the same weight, only one thirtieth part. Neglect of the first of these corrections would occasion error in the weight of the gallon to the extent of nearly three grains (2.88 gr.) ; while, in the case of the second, the error would ordinarily be less than one grain ; and would amount to only about two, on the supposition, just made, of the extreme barometric variation of an entire inch.

Supposing, however, that the regulating weight of the measures of capacity is *not* the weight in the air, but the weight *in vacuo*, of a cubic inch of water of maximum density, it is the *total* weight of the air displaced by the water, less that of the air displaced by the counterpoises, which should be allowed for in the adjustments. Such a regulating weight appears, as shown in the preceding note, to have been selected by Mr. HASSLER ; yet his formulas for correction are constructed as if this weight had been determined by weighing in the air under the normal pressure at the temperature of 62° F. Hence would follow the theoretic consequence, that the bushel and gallon, at the standard temperature, must be slightly above their legal capacity. But inasmuch as this regulating weight, derived as we have seen from the kilogramme, is below that deduced from the British experiments as being the weight under normal pressure in the air, there seems to be a probability that the error of the standard measures of capacity of the United States, if any exists, is on the side of deficiency rather than of excess.

Note 2.—REËXAMINATION OF CAPTAIN KATER'S DETERMINATION OF THE WEIGHT OF A GIVEN VOLUME OF WATER.

In recomputing Captain KATER's results, referred to in the preceding note, employing his own data, some slight, though not very important, discrepancies are observed. The details of the recomputation are here presented.

I. FOR THE CUBE:

Temperature of air during the weighing. = 62° F.

  " " water " " = 60°2 F.

Height of barometric column during the weighing.......................... = 29 inches.

Specific gravity of air at 62° F. and 30 inches barometer................... = $\frac{1}{834}$

Specific gravity of weights employed.... = 8.5.

Density of solid to density at 62° F. as unity....................... = 1.0000567.

Density of water to density at 62° F. as unity........................... = 1.00017.

Volume of solid at 62° F.............. = 124.1969 cu. in.

Apparent weight of water displaced.... = 31,381.79 grains.

With these data, the following are the results obtained as compared with those originally deduced by KATER himself from the same.

1. Correction for displacement:

KATER'S RESULT.	RECOMPUTATION.
$31,381.79 \cdot \frac{1}{834} \cdot \frac{29}{30} \cdot \frac{7.5}{8.5} = 32.000.$	32.094.

2. Weight *in vacuo* of water displaced:

KATER'S RESULT.	RECOMPUTATION.
$31,381.79 + 32.00 = 31,413.79.$	$31,381.79 + 32.094 = 31,413.884.$

3. Reduction to the standard temperature:

KATER'S RESULT.   RECOMPUTATION.

$31{,}413.79 \cdot \dfrac{1.0000567}{1.00017} = 31410.24.$ $31{,}413.884 \cdot \dfrac{1.0000567}{1.00017} = 31{,}410.3264.$

4. Weight *in vacuo* of cubic inch of water at 62° F.:

KATER'S RESULT.   RECOMPUTATION.

$\dfrac{31{,}410.24}{124.1969} = 252.907.$ $\dfrac{31{,}410.3264}{124.1969} = 252.90754.$

II. FOR THE CYLINDER:

In this case, the temperature of the air at the weighing, and therefore, also, its specific gravity, remain the same as before. The stand of the barometer is likewise the same, and the same counterpoise weights continue to be used. The other conditions are different, as follows:

Temperature of the water during the weigh-
ing .... $= 60°5$ F.
Density of solid to density at 62° F. .... $= 1.000047.$
Density of water to density at 62° F.... $= 1.0001456.$
Volume of solid at 62° F. .... $= 75{,}2398$ cu. in.
Apparent weight of water displaced..... $= 19{,}006.83$ grains.

From which we obtain,

1. Correction for displacement :

KATER'S RESULT.   RECOMPUTATION.

$19{,}006.83 \cdot \dfrac{1}{834} \cdot \dfrac{29}{30} \cdot \dfrac{7.5}{8.5} = 19.43.$ $19.4385.$

2. Weight *in vacuo*, of water displaced:

KATER'S RESULT.   RECOMPUTATION.

$19{,}00683 + 19.43 = 19{,}026{,}26.$ $19{,}006.83 + 19.4385 = 19{,}026.2685.$

3. Reduction to standard temperature:

KATER'S RESULT.   RECOMPUTATION,

$19{,}026.26 \cdot \dfrac{1.000047}{1.000145} = 19{,}024.46.$ $19{,}026.2685 \cdot \dfrac{1.000047}{1.0001456} = 19{,}024.4123.$

4. Weight *in vacuo* of cubic inch of water:

KATER'S RESULT.

$$\frac{19,024.46}{75.2398} = 252.851.$$

RECOMPUTATION.

$$\frac{19,024.4123}{75.2398} = 252.85035.$$

III. FOR THE SPHERE:

The conditions in this experiment are all changed, excepting the specific gravity of the counterpoise weights. The rest are as follows:

Temperature of the air during the weighing .... $= 67°$ F.

Temperature of the water during the
weighing .... $= 66°$ F.

Height of barometric column .... $= 29.74$ inches.

Specific gravity of air at 67° F. .... $= \frac{1}{843}$.

Density of solid to density at 62° F..... $= 0.999874.$

Density of water to density at 62° F.... $= 0.99958.$

Volume of solid at 62° F. .... $= 113.5264$ cu. in.

Apparent weight of water displaced .... $= 28,673.51$ grains.

Whence we deduce—

1. Correction for displacement:

KATER'S RESULT.

$$28,673.51 \cdot \frac{1}{843} \cdot \frac{29 \ 74}{30} \cdot \frac{7.5}{8.5} = 29.72.$$

RECOMPUTATION.

$$29.7519.$$

2. Weight *in vacuo* of water displaced:

KATER'S RESULT.

$$28,673.51 + 29 \ 72 = 28,703.23.$$

RECOMPUTATION.

$$28,673.51 + 29.7519 = 28,703.2619.$$

3. Reduction to standard temperature:

KATER'S RESULT.

$$28,703.23 \cdot \frac{0.999874}{0.99958} = 28,711.66.$$

RECOMPUTATION.

$$28,703.2619 \cdot \frac{0.999874}{0.99958} = 28,711.6992.$$

4. Weight *in vacuo* of cubic inch of water:

KATER'S RESULT.

$$\frac{28,711.66}{113.5264} = 252.907.$$

RECOMPUTATION.

$$\frac{28,711.6992}{113.5264} = 252.90768.$$

	KATER'S RESULT.	RECOMPUTATION.
Resulting mean........................ ...	252.888	252.8885
Mean in parliamentary grains............	252.722	252.7225
Whence weight in air, 62° F. and 30 in. bar.	252.456	252.4551

In the final result the discrepancies which appear in the process of calculation nearly balance each other.

The computations of Captain KATER, here reviewed, form an appendix to the third report of the parliamentary committee of 1818. In the report itself, the weight of the cubic inch of water *in vacuo* at 62° F. is given as 252.754, and the Appendix is referred to as authority. The discrepancy between the two statements is not explained , but the statement of the report furnishes a reason if not a justification for the adoption in the British statute of 1824 (5 GEORGE IV., c. 74), of the value 252.458 as that of the weight of a cubic inch of water in the air, both water and air being at the temperature of 62° F., and the barometer standing at thirty inches—a value upon which Mr. HASSLER has based his reductions to the temperature of maximum density in his report of 1832 ; though, as is shown in the NOTE foregoing, he set aside the value found by these reductions in favor of that derived from his authentic brass kilogramme, in fixing the weights which determine the standard measures of capacity of the United States. Mr. HASSLER's reductions, here referred to, take no account of the different specific gravities of the air at the two temperatures considered ; a circumstance which to a certain slight extent vitiates the result, whether the reduced weight be referred to air at 39°1 F., or to air at 62° F.

The specific gravities of air employed by Captain KATER in the calculation above reviewed are certainly too small. It is worth while, therefore, to repeat the operations with new constants. The following values for the weight in grains of a cubic inch of dry air at 60° F. of temperature, and under a barometric pressure of thirty inches, are given by the several authorities named :

Biot and Arago............................ 0.31074
Dumas and Boussingault 0.31086
Regnault................................. 0.30938
Prout.................................... 0.31017

Mean........................... 0.3102675

If we reduce to 62°, supposing the pressure constant, and assuming the absolute zero to be at 461° below 0° F., we shall obtain the result 0.30908093 grs. ; and taking the cubic inch of water *in vacuo* at the same temperature to be 252.722 as above, we shall obtain a specific gravity of 0.0012230076, or $\frac{1}{817.656}$.

We may presume the weighings of the experimenters whose determinations are here cited, to have been made without allowance for the displacement of air by the counterpoise weights. The displacement of the air-vessel employed in the experiments would be the same, of course, before and after exhaustion.* The allowance for the counterpoises, if made, will slightly reduce the observed weight ; and will reduce, of course, the resulting specific gravity. This allowance is made here accordingly, that nothing may be omitted which tends to diminish the difference between the specific gravities deduced from these weights and those employed by KATER. The counterpoises are supposed to be of brass, having the specific gravity of those used in the British experiments, viz., 8.5. The weight, after the correction thus indicated, becomes 0.309036458 grs. ; and the resulting specific gravity is 0.0012228315 $= \frac{1}{817.774}$.

This is on supposition that the air is dry. In RANKINE's *Manual of the Steam Engine*, there are given tables of the pressure and weight per cubic foot of watery vapor from 32° F.

---

* In order to eliminate entirely the effect of displacement, in experiments on the weight of air and gases, the expedient was employed by REGNAULT and PROUT, of suspending to the opposite scale of the balance, a second vessel exactly similar in form and equal in weight to that containing the gas to be weighed.

upward. It is reasonable to assume the average depression of
the dew point to be not less than six degrees. If, therefore, to
the weight of a cubic inch of dry air as already found, we add
the weight of a cubic inch of watery vapor, supposed saturated at
a temperature six degrees below, and if from the same we sub-
tract a fraction of its total weight indicated by the ratio of the
tension of the vapor to the total tension, the result should be the
ordinary weight of a cubic inch of air at the standard temperature and pressure. These operations having been performed,
we find at length the weight at 62° F., proper for our purposes, to be 0.307386258 grains ; giving a specific gravity of
$0.001216302 = \frac{1}{822.164.}$ A similar computation gives the weight
of a cubic inch of air at 67° F. and 30 inches of pressure as
0.304150644 grs., with a specific gravity to water at 62° F., of
$0.001203498 = \frac{1}{830.91.}$

In the calculations given below, these values are employed ;
and the densities of water at different temperatures, instead of
being taken from GILPIN, are derived from the more recent
determinations of KOPP. The constants for the expansion and
contraction of brass remain as stated by KATER. The temperature of the maximum density of water as given by HASSLER
(39°.83 F. = 4°.35 C.) is doubtless too high. It is given by
CHISHOLM, on the authority of PLAYFAIR and JOULE, (*Second
Report of the Royal Standards Commission*, 1869, Appendix VI.)
at 3°.945 C. = 39°.101 F. ; and inasmuch as the variation of
density is almost insensible near that point, we may assume it
for the purposes of this calculation at 4° C = 39°.2 F.

The following then are the coëfficients of volume of water at
the several temperatures named, as taken directly or deduced
by interpolation from the tables of KOPP.

At 32° F.= 0°.0 C 1.000000   At 60°.5 F.= 15°.83 C. 1.000820
" 39.2 " = 4.0 " 0.999877   " 62.0 " = 16.67 " 1.000953
" 60.2 " = 15.662" 1.000795   " 66.0 " = 18.89 " 1.001350

For the dilatation and contraction of brass, if we take the density at 62° F. as unity, we shall have the following coefficients.

For 60°.2 F........ 1.00005670   For 62° F. 1.00000000
" 60.5 " .... 1.00004725   " 66 " 0.99987400

Putting now

W $=$ the apparent weight of the water displaced.

W' $=$ the weight after correction for displacement.

W" $=$ the corrected and reduced weight, being the weight at 62° F. *in vacuo.*

S $=$ specific gravity of dry air at 62° F. to water at 62° F.

c $=$ correction of S for humidity. It is the weight per cubic inch of saturated aqueous vapor at 6° F. below the temperature of observation, less the proportion of the total weight per cubic inch of vapor and dry air, expressed by the ratio of tension of vapor to joint tension of vapor and air,—divided by the weight per cubic inch of water at 62° F. *in vacuo.*

s $=$ specific gravity at 62° F. of body weighed to water at 62° F.

T $=$ British standard temperature (62° F.), referred to the absolute zero ($-461°$ F).

τ $=$ temperature of the air at the weighing, referred to same.

P $=$ standard barometric pressure.

p $=$ barometric pressure at the weighing.

V $=$ coefficient of volume of water at 62° F., from table.

v $=$ " " " " at temperature of experiment.

D $=$ density of brass at 62° F.

d $=$ " " at temperature of experiment.

C $=$ correction for displacement.

The correction for displacement, to be added to W, will be found by the formula,

$$C = W \cdot \left(S \cdot \frac{T}{\tau} + c \right) \cdot \frac{v}{V} \cdot \frac{P}{P} \cdot \frac{s\,D\,v - D\,V}{s\,D\,v} \cdot {}^*$$

This correction having been added to W, the sum W′ will represent the weight *in vacuo* of the volume of water actually displaced by the solid, with the density belonging to the water at the temperature of the experiment. This volume is then to be reduced to the volume at 62° F., by means of the coefficients of density of brass given above; and the density, to the density at 62°, by means of those of volume for water, in accordance with the following formula,

$$W'' = W' \cdot \frac{v}{V} \cdot \frac{D}{D}.$$

Performing the operations and putting $C_A, C_B, C_C, W_A, W_B, W_C$, to represent the several corrections, and the several weights as corrected, we find,

$$C_A = 32.5513. \quad C = 19.7157. \quad C_C = 30.5180.$$

And

$$W_A = 31,414.3413. \quad W_B = 19,026.5157. \quad W_C = 28,714.0280.$$

These, further reduced by the second formula, give

$$W_1 = 31,411.0808. \quad W_2 = 19,024.8829. \quad W_3 = 28,711.7140.$$

Whence, dividing by the measured bulks of the solids immersed, there are obtained the three values for the weight of the cubic inch of water *in vacuo* at 62° F., and 30 inches of barometric pressure, 252.9136 grs., 252.8566 grs., and 252.9959 grs. ; giving a mean of 252.9220 grs. This weight being expressed in the grains of Sir GEORGE SHUCKBURGH, must be reduced to the parliamentary standard, by multiplying by the coefficient 0.9993432 which gives as a final result, 252.75965 grs. standard.

There is reason, therefore, to think that the reductions made by Captain KATER from the experiments of Sir GEORGE SHUCK-

---

* This last factor $\frac{s\,D\,v - D\,V}{s\,D\,v}$ is put into this form only because it is what the rigid theoretic truth requires. Practically the specific gravity, s, may be regarded as invariable; and in place of this factor, may be written $\frac{s-1}{s}$.

BURGH, after the remeasurement of the solid bodies employed in those experiments had been carefully made by himself, give a result sensibly though slightly too small, the difference amounting to 0.038 grs. nearly.

If, from the weight just found, we deduct the proper correction for displacement in weighing in the air, at the British standard temperature and pressure, with brass weights of the specific gravity of 8.5, we shall obtain for the cubic inch of water under these conditions, the value 252.48843 grs. instead of 252.456, the result found by KATER.

Reducing the weight *in vacuo* found above to the temperature of maximum density, by means of the coefficients of volume from the table of KOPP, we find for the cubic inch of water under these conditions, the weight 253.0286. KATER's value 252.722, reduced by means of the coefficient of Prof. MILLER employed in the report of 1869 of the Standards Commission of Great Britain, gives 253.00452 grs. The corresponding value, as deduced from the kilogramme, according to MILLER, we have seen to be only 252.87718 grs.* The difference is 0.12734 grs.; which, multiplied by 61.02705152, the number of cubic inches in a cubic decimetre, gives 7.77118 grs., the amount by which the platinum prototype kilogramme of the Archives would appear, from this determination, to be too light. If the comparison be made with the value found in the foregoing computation, viz. :

---

* As there are three determinations of the weight of the kilogramme in British standard grains, all entitled to respect, viz. : (taken in the order of publication) those of HASSLER (1832), KUPFFER (1841), and MILLER (1844); and also three determinations of the metre in British standard inches, and therefore of the cubic decimetre, or *litre*, in British standard cubic inches, similarly deserving of consideration, viz. : (taken likewise in the order of publication) those of KATER (1821), HASSLER (1832), and CLARKE (1866), it follows that, by combining these data by pairs, we may obtain nine values, slightly differing from each other, for the weight *in vacuo* of the cubic inch of water at maximum density, as derived from the kilogramme. It may be interesting to have all these under the eye in a single group. The data are as follows :

	HASSLER.	KUPFFER.	MILLER.
Kilogramme (grains)...	15,433.1669	15432.36186	15432.34874
	KATER.	HASSLER.	CLARKE.
Metre (inches)	39.37079	39.3810327	39.370432
Litre (cu. in.).........	61.02705152	61.07469428	61.02499296

Taking, now, the initial letters of the names of these authorities to indicate the several

253.0286 grs., the discrepancy becomes as great as 9.24072 grains.*

Returning to the weight of the cubic inch of water *in vacuo* at maximum density, as found by our own reductions above, we shall obtain the weight in air at the corresponding temperature, by correcting, as in previous computations, for displacement. From the weight of the cubic inch of dry air at 62° F. and at the normal pressure, we deduce the weight at 39°.1 F. under the same pressure, by the ordinary methods ; which weight we find to be 0.323187315 grs. Correcting then, as in the previous cases, for humidity, we obtain the reduced weight 0.321240544 grs. And multiplying this by the constant 0.882353, which expresses the displacement by the water (taken as unity) less the displacement by the brass weights, we arrive finally at the correction 0.2834476 grs., which is to be subtracted from the weight *in vacuo* as found above, viz., 252.0286 grs. ; leaving finally

---

values of the kilogramme and of the litre, and using the large capitals for the first, and the small capitals for the second, we have these combinations :

WEIGHT, Cu. in. water, *in vacuo.*	WEIGHT, Cu. in. water, *in vacuo.*	WEIGHT, Cu. in. water, *in vacuo.*
H + K = 252.89059	K + K = 252.87736	M + K = 252.87718
H + H = 252.69330 (a)	K + H = 252.68033	M + H = 252.67992
H + c = 252.89912	K + c = 252.88593	M + c = 252.88593

(a) This falls short of the value given by Mr. HASSLER himself (252.6934) by one ten thousandth of a grain. The difference arises from the fact that, in his value of the litre, the fifth decimal figure is 6 instead of 9, as, from his value of the metre, it should be.

The largest of these values is obtained from HASSLER's kilogramme and CLARKE's metre ; the smallest from MILLER's kilogramme and HASSLER's metre. The mean of the whole is 252.81885 grains ; which is less than the result of the computation in the text by 0.20975 grs. The mean of the litres is 61.04224625 cubic inches ; and if the difference just found be multiplied by this, the kilogramme of the Archives will appear to be below the theoretic weight as determined by the British experiments, to the extent of 12.80361 grains. Rejecting the values dependent on the metre of Mr. HASSLER, we shall have a mean of 252.88451 grains, for the cubic inch of water, and a mean litre of 61.02602224 cubic inches. This mean weight falls below the computation in the text by 0.14409 grs. ; indicating that the prototype kilogramme is 8.79324 grains light. But if the comparison be made with the weight deduced from the report of the Standards Commission, as given above, viz., 253.00452, the apparent deficiency per cubic inch is only 0.12001 grs., and the inferred deficiency of weight of the prototype kilogramme is reduced to 7.32373 grains.

* In the Report above referred to, of the Standards Commission, to the British Parliament, made in 1869, the Warden of the Standards, H. W. CHISHOLM, Esq., singularly enough inverts this statement, and says that the standard kilogramme is "*in excess of its theoretical weight.*" According to his computation, the difference is 7.76247 grains.

252.7452 grains as the weight of the cubic inch of water in air at 39°.1 F., and under thirty inches of barometric pressure. This value is identical with that of VAN DER TOORN, given by HASSLER, in his report of 1832, as "the English determination," excepting only the slight difference of a unit in the fourth decimal place. The coincidence is remarkable, in view of the widely different processes by which the two have been reached.

Dividing the weight of the cubic inch of humid air at 39°.1 F., as found above, by the weight of the cubic inch of water at maximum density, we shall have, as the specific gravity of air under these circumstances, $0.001269582 = \frac{1}{787.66}$. It is this specific gravity which ought to be employed in the corrections for barometric fluctuations during the adjustment of measures of capacity, in case the weight of the standard cubic inch of water according to which their capacities are computed, has been determined on supposition that the air as well as the water is at the temperature of maximum density of water ; and that the adjustments are made in an atmosphere of the same temperature. On this supposition, the weight 252.7452 grains, just found, is the proper standard weight of the cubic inch of water ; but if this standard weight is supposed to be taken with water at the maximum density and air at 62° F., its value ought to be stated at 252.7574 grains. The difference, as is pointed out in the preceding note, will affect the weight of the gallon to the extent of nearly three grains.

APPENDIX C.

NOTE ON BRITISH LEGISLATION IN REGARD TO THE METRIC SYSTEM.

(REFERRED TO ON PAGE 49.)

The history of British legislation in regard to the Metric System of Weights and Measures having been but imperfectly given in the text of the foregoing address, it has been thought proper to append the following succinct account of it, extracted from a paper by H. W. CHISHOLM, Esq., Warden of the Standards of Great Britain, which was published in connection with the "Second Report of the Standards Commission," in 1869.

Mr. CHISHOLM says :

"The last evidence which I shall bring before the Commission of the progress of public opinion in this country as to the introduction of the Metric System, is that of the proceedings in Parliament. And first, as to the Weights and Measures Committee of 1862. That Committee was composed of members who formed opinions entirely favorable to the introduction of the Metric System, and the terms in which their recommendations are framed in their report deserve particular notice.

"They 'arrived at a unanimous conclusion that the best course is cautiously but steadily to introduce' ' the Metric System into this country,' and they recommended—

"1. That the use of the Metric System be rendered legal. No compulsory measures should be resorted to until they are sanctioned by the general conviction of the public.

"2. That a Department of Weights and Measures be established in connection with the Board of Trade. It would thus become subordinate to the Government and responsible to Parliament.   To it should be intrusted the conservation and verification of the standards, the superintendence of inspectors, and the general duties incident to such a department.   It should also take such measures as may from time to time promote the use and extend the knowledge of the Metric System in the Departments of Government and among the people.

" 3. The Government should sanction the use of the Metric System (together with our present one) in the levying of the customs duties, thus familiarizing it among our merchants and manufacturers, and giving facilities to foreign traders in their dealings with this country.   Its use, combined with that of our own system in government contracts, has also been suggested.

" 4. The Metric System should form one of the subjects of examination in the competitive examination of the Civil Service.

" 5. The gramme should be used as a weight for foreign letters and books at the Post Office.

" 6. The Committee of Council on Education should require the Metric System to be taught (as might easily be done by means of tables and diagrams) in all schools receiving grants of public money.

" 7. In the public statistics of the country, quantities should be expressed in terms of the Metric System, in juxtaposition with those of our own, as suggested by the International Statistical Congress.

" 8. In private bills before Parliament the use of the Metric System should be allowed.

" 9. The only weights and measures in use should be the Metric and Imperial, until the Metric System has been generally adopted.

*   *   *   *   *   *   *   *   *

" As a consequence of the report of the committee of 1862, a bill to introduce the Metric System was introduced in the following session, founded on their recommendations ; but it also

contained the provision that it should be made compulsory at
the expiration of a period to be fixed by the bill.   A period of
three years was inserted in the draft of the bill, although it was
stated by its promoters that a longer period might be fixed, as
might be judged expedient or necessary.   The compulsory provision was objected to by the Government ; but, on a division,
the second reading was carried by 110 against 75 on July 1st,
1863.   The lateness of the session prevented the bill being then
proceeded with; but it was re-introduced in 1864 as a permissive bill only.   The second reading on 10th March was not
opposed by the Government, and was carried on a division by
90 against 38.   In committee on 4th May, the Government
objected to the provisions for legalizing Metric standards and
weights and measures, there being no arrangements for providing them, and proposed themselves to substitute a new bill,
rendering the Metric System permissive in this country so far
as related to contracts.   This proposal was acceded to, and the
bill finally passed as the Metric Act of 1864, though not until
after a division upon the second reading in the House of Lords
on 21st July, of 34 against 23.

" In 1868, another bill was brought in, to repeal the act of
1864, and introduce the Metric System compulsorily, after a
period to be fixed in the bill, which period was left to be inserted
by the Government.   Upon the second reading of the bill, on
13th May, 1868, it was not opposed by the Government, on the
understanding that they did not object to its principle, but that
further information was required before the question could be
properly dealt with, and was then being obtained by the Standards Commission, and that the bill should not be proceeded with
until the report of the commission had been received.   This understanding was assented to by the promoters of the bill, and
the bill was read a second time after a division of 217 against 65.

" The time has now arrived when all the evidence that appears
to be required to form their judgment has been placed before

the Standards Commission, and when the whole question has become ripe for their opinion to be given upon it. Upon full consideration of all the circumstances, and after the adoption of the Metric System by so many other countries, and the influential and increasing support it has met with here, I am myself led to the irresistible conviction that the full introduction and legalization of the Metric System in the United Kingdom can no longer be delayed. Considering the extent and importance of our commercial and scientific intercourse with so many nations who have adopted the Metric System of weights and measures, and that it has now become really a widely extended international system, it appears to me that this country can no longer isolate herself from other countries in her system of weights and measures, and refuse to adopt that common method of computing the quantities of all merchandise and other articles passing between them, which so many other nations have already accepted."

The views here expressed by Mr. Chisholm were approved by the commission,* who, in the report above referred to, submitted the following among their recommendations to Parliament:

" (1.) Considering the information which has been laid before the Commission—

"Of the great increase during late years of international communication, especially in relation to trade and commerce;

"Of the general adoption of the Metric System of weights and measures in many countries, both in Europe and other parts of the world, and more recently in the North German Confederation and in the United States of America;

"Of the progress of public opinion in this country in favor of the Metric System as a uniform international system of weights and measures;

* The commission was composed as follows ;—GEORGE BIDDELL AIRY (Astronomer Royal), Chairman ; Lord COLCHESTER ; Mr. STEPHEN CAVE ; Mr. JOHN GEORGE SHAW LEFEVRE ; General EDWARD SABINE ; Sir THOMAS GRAHAM (Master of the Mint); Mr. W. H. MILLER (of the Bank of England), and Mr. H. W. CHISHOLM (Warden of the Standards).

" And of the increasing use of the Metric System in scientific researches, and in the practice of accurate chemistry and engineering construction;

" We are of opinion that the time has now arrived when the law should provide, and facilities be afforded by the Government, for the introduction and use of Metric weights and measures in the United Kingdom.

*   *   *   *   *   *   *   *   *
*   *   *   *   *   *   *   *   *

" (2.) Considering the advantage of adopting in an international system not only of uniform weights and measures, but also uniform names ; and that although there may be well founded objections to the inconvenient length and occasional similarity, both to the eye and ear, of the French nomenclature, yet it is probable that these names will become familiar by custom, and obtain popular abbreviations ;

" We think that the French nomenclature, as well as decimal scale of the Metric System, should be introduced in this country.

" (3) Considering—

" That there is no immediate cause, requiring a general change in the existing system of legal weights and measures of the country for the purposes of *internal* trade ;

" That the statutable values of the fundamental imperial units are adopted in use without the slightest variation throughout the whole of the British Isles ;

" That the primary Imperial Standards are as perfect as can be made by modern skill and science, and that the whole series of official standards are now most accurately verified in relation to the primary standards.   *   *   *   *   *
*   *   *   *   *   *   *   *   *
*   *   *   *   *   *   *   *   *

" We are of opinion that the *general* introduction of the Metric System should be permissive only, and not made compulsory by law after any period to be now specified, so far as relates to the use of Metric weights and measures for weighing and measuring goods for sale or conveyance.   *   *   *   *   *   *

" (5) We are of opinion that it is expedient that customs duties should be allowed to be levied by Metric weight and measure, as well as by Imperial weight and measure ; that the use of the Metric System concurrently with the Imperial System, should be adopted by other public departments, especially the Post Office, and in the publication of the principal results of the Statistics of the Board of Trade, as well as for the admeasurement and registration of the tonnage of shipping ;

" (6) And that mural standards of the Metric System, as well as of the Imperial System, be exhibited in public places.

" (7) Considering—

" That the Metric System, as adopted in other countries, includes the relation of coinage to weights and measures, particularly in its uniform decimal scale ;

" And that the advantages of the introduction of the Metric System into this country, as an international system of weights and measures, would be much increased by establishing a corresponding international system of coinage, in regard to a unit and to a decimal scale ;

" We are of opinion that, even if the difficulties of establishing an international unit of coinage cannot be at present overcome, yet the decimalization of our system of coinage, which is in the power of the Government, would be very useful to the public."

These very judicious recommendations of the Standards Commission are believed to have met with the general approval of the thinking people of Great Britain. As yet however the national legislature has not taken the decisive step which is evidently inevitable at no distant day, of adopting the Metric System of Weights and Measures as the only system to be legally used in the United Kingdom. During the recent session of Parliament (1871), a bill was introduced having this object in view, which, being opposed by the ministry, was lost on a very close division. The following extracts from a report of the debate on the subject in the House of Commons, are derived from the London *Guardian*, of August 2, 1871:

" On Wednesday [July 26], Mr. J. B. SMITH moved the second reading of his bill for the compulsory introduction of the metric or decimal system. The honorable member, after dwelling on the curious jumble of weights and measures which were still in vogue amongst us, said that the act of 1864 was so badly drawn by the Board of Trade that in the opinion of the law officers the Metric System was legal, and yet that if metric weights and measures were found on a man's premises he was liable to prosecution. The Metric System was in compulsory use amongst 200,000,000 people, and in permissive use amongst 200,000,000 more. In fact, sixty-six per cent. of our exports went to countries using that system; and its introduction amongst us would save infinite trouble, first, in education, and afterwards in the operations of commerce. Moreover, Sir ROWLAND HILL had further stated that we were losing six per cent. on our postage with France, and seventeen per cent. on our postage with Prussia, in consequence of a want of identity in weights and measures.

" The bill was supported by forty-three Associated Chambers of Commerce and Agriculture, by Farmers' Clubs, Workingmen's Associations, and many scientific bodies. It was supported by the representatives of the largest constituencies in the kingdom —Manchester, Liverpool, Glasgow, Leeds, Birmingham, North Staffordshire, South Leicestershire, and South Norfolk. It was opposed by the member for Cambridge University, on the ground that the French unit was not a proper unit ; by the Astronomer Royal and Sir J. HERSCHEL. For the Metric System however they had the authority of three gentlemen who possessed the most extensive scientific knowledge, combined with the greatest business knowledge of this or any other age—Sir WILLIAM ARMSTRONG, Sir JOSEPH WHITWORTH, and Sir WILLIAM FAIRBAIRN.

"In a word,. he asked for what the English barons had demanded 700 years ago—namely, that there should be one weight and one measure throughout the land, but he asked that that weight and that measure should be identical with those of other nations, so that, like a common language, there should be but one weight and one measure throughout the world."

The passage of the bill was resisted on the usual grounds, and on some that are unusual.   For instance—

"Mr. BERESFORD HOPE, in moving the rejection of the bill, said that, while he was in favor of uniformity, he desired to have English weights and measures, and not French ones.   The changes proposed by Mr. SMITH would overthrow all our long-established habits and customs—nay our very proverbs ; for no one would be able hereafter to talk about 'giving an inch and taking an ell'—(a laugh):—"

This latter argument of Mr. BERESFORD HOPE is rather more witty than weighty, but it is less true than either : for though the ell (which after all is a French measure) has ceased to be used in England for a century or two, the proverb seems to be still as lively as ever.   As for stigmatizing the Metric System as *French*, it was never quite just to do so, and it is at present altogether absurd ; since this system has become the system of more than half the civilized world.

"The rejection of the bill was seconded by Mr. STEVENSON, and was supported by Mr. SCOURFIELD, by Alderman LAWRENCE, who pointed out that the adoption of the *litre* would diminish the poor man's pint of beer without any proportionate diminution of price, [a surprisingly knowing alderman this must have been] and by Mr. FOTHERGILL.   Mr. HENLEY pointed out the inconvenience of having to go to a foreign capital for standards which might at any moment be melted down in a general conflagration of the city.   If agriculturists wanted a uniform measure, why on earth did they not adopt the Imperial bushel ?"

The honorable gentleman proceeded, "When the Government

could say there was a reasonable certainty that all other nations would adopt the same uniform system, it would be time enough to ask our people to bear such a tax, and to make a change which would create great internal confusion. [Let us suppose "all other nations" to wait in like manner for "all others"—what would probably be the rate of progress?] They knew what had happened in France; how Frenchmen first began to cut off their king's head [it is commonly supposed they finished], declared there was no God, set up a social evil as an object of worship, and cast aside all the ancient provinces and limits of the land. There was also an enormous change of property, almost to the extent of being universal; and when that was done, they set about measuring the world, and enforced that unit which people were so fond of throughout France, where they had a vast number of different measures in every province. He would be glad to hear how long it took, with all the despotism they had in France, before the old system of weights and measures was fairly got rid of. That was a matter upon which the government ought to give them some information." [Here is a gentleman who evidently ought to read the reports of his own "Standards Commission."]

"The bill was supported by Sir C. ADDERLEY, by Mr. BAINES, by Mr. CLARE READ (who said he had bought and sold in Oxfordshire, Norfolk, and Pembrokeshire by different weights for different articles; and in Shropshire, where they actually had different weights for different market days), by Mr. PELL, by Col. SYKES, and by Mr. ILLINGWORTH. Mr. CHICHESTER FORTESCUE, the President of the Board of Trade, said that this was the first time the issue had been really raised in Parliament; for the purport of the measure of 1864 was much misunderstood. It was true that it bore the title of an act to legalize the use of metric weights and measures; but all it did was to render contracts under the metric system legal, which they were not before; but it by no means legalized the use of those weights and measures in this country. * * * In his opinion the general feeling of this country was not at present such as would justify the compulsory introduction of the metric system. Under all the circumstances it would be better to postpone the consideration of the matter, and leave it to the government to introduce a complete system next session." [Here was a *quasi* promise on the part of the ministry to favor the introduction of the system at a later day, and possibly with less abruptness. Perhaps it would have been better to accept the overture, but the friends of the bill thought otherwise.]

"Mr. EASTWICK advised Mr. SMITH to accept this offer: but the honorable member insisted on a division, and was beaten by eighty-two to seventy-seven."

In connection with the foregoing account of the home legislation of Great Britain, the following statement, derived from a little volume recently published, (June, 1871), by Professor LEONE LEVI, of London, Honorary Secretary of the Metric Committee of the British Association for the Advancement of Science, will be found interesting as it respects the simultaneous movement of public opinion, and its effect upon metrological reform, in British India:

"The necessity," says Prof. LEVI, "of providing for uniformity of weights and measures in India, engaged the attention of the Indian Government frequently since 1837. On the 13th of May, 1863, the Government of Madras suggested the importance of steps being taken to reform the various systems of weights and measures, and on the 20th June, 1864, a resolution was passed by the Government of India, recommending the appointment of local committees in each Presidency, to deliberate and report on the whole matter. At Madras and Bombay, in the north-western provinces, and the Punjaub, the committees reported in favor of a scheme based upon the Imperial system. But the Bengal Government proposed the gradual but finally complete adoption of the Metric System, which recommendation received the support of the Bengal Government. On the 1st October, 1867, Colonel STRACHEY, President of the Central Committee, published a memorandum in support of the proposal of the Bengal Committee, and included in it a draft of a bill on the subject. The majority of his colleagues were, however, opposed to his measure, and Colonel STRACHEY retired from the Committee, which afterwards reported against the adoption of the Metric System, and recommended the scheme based on the Imperial. At about this time a deputation of the Metric Committee of the British Association and the Council of the International Decimal Association, waited on Sir STAFFORD NORTHCOTE, urging the superior claims of the Metric System for India, which was communicated by him to the Indian Government. And on the 5th September, 1868, a decisive Minute was made by the Commander-in-Chief, Sir W. R. MANSFIELD, in favor of the Metric System with regard to weights. Sir JOHN LAURENCE approved of the Minute, and so did the Hon. Sir R. TEMPLE. The matter was then referred by the Governor-General in Council to the Duke of Argyle, Secretary for India, who consulted the Board of Trade on the subject. The approbation of the Board having been duly obtained, an act was passed in 1870 to regulate the weights and measures in British India. The act constituted the primary standard of weight, the *Ser*, of the same weight as the kilogramme, and the primary standard of length, the metre; and it empowered the Governor-General in Council to cause such new weights and measures to be used by any government office or municipal body or railway company; and to require that, after a date to be fixed, all or any of the weights and measures of capacity aforesaid shall in every district be the basis of the dealings and contracts of all persons engaged in any specified business or trade. It also provided that after the date fixed in any such notification, all dealings, or contracts made by the officers, bodies, companies, or persons so mentioned, for any work done or goods sold by weight or measure, shall, in the absence of a special agreement to the contrary, be deemed to be had or made, according to the weights and measures directed in such notification to be used.

APPENDIX D.

NOTE ON THE EXTENT TO WHICH THE METRIC SYSTEM HAS BEEN ALREADY ADOPTED.

(REFERRED TO ON PAGE 51.)

The statements of the foregoing address as to the populations among whom the metric system has been wholly or partially legalized, though made upon less full information than has since been collected, have been generally allowed to stand as they were originally written. The object of this note is to present a more complete view of the facts, (so far as ascertained up to December, 1871,) than could be introduced into the body of the address, without too largely changing its form. In the follqwing tables are embraced, by classes, all the nations whose legislation is known to have favored this system; distinction being made between those which have adopted the system in full; those which have established simple relations of commensurability between it and their own; those which, still retaining their own unit bases, have adopted metric ratios for the derivative denominations; and, finally, those which have made the use of the system optional but not compulsory.

I.—*Peoples adopting the Metric System in full.*

STATE.	YEAR.	POPULATION.
France...........................	1866	38,067,064
French Colonies	1866	2,921,000
Holland...........................	1868	3,638,467
Dutch Colonies.	1868	22,453,000
Belgium	1866	4,839,094
Spain.............	1868	16,642,000

TABLE I.—*Continued.*

STATE.	YEAR.	POPULATION.
Spanish Colonies...............	1868	2,030,000
Portugal.......................	1863	4,349,000
Italy......!	1868	25,527,000
North German Confederation......	1867	29,910,517
Greece.........................	1864	1,348,522
Roumania......	1867	4,605,000
British India...................	1866	150,767,851
Mexico.........................	1865	8,218,080
New Granada....................	....	2,800,000
Ecuador........................	1858	1,040,000
Peru...........................	....	3,374,000
Brazil..........................	1867	9,858,000
Uruguay	....	387,000
Argentine Confederation..........	1869	1,736,000
Chili..........................	1868	1,908,000
Total.....................	....	336,419,595

II.—*Peoples adopting Metric Values.*

STATE.	YEAR.	POPULATION.
Würtemberg....................	1867	1,778,396
Bavaria........................	1867	4,824,000
Baden	1867	1,438,000
Hesse..........................	1852	854,319
Switzerland	1860	2,510,494
Denmark.......................	1850	2,413,000
Austria	1867	34,861,000
Turkey.........................	....	35,360,000
Total.....................	....	84,039,209

III.—*Countries in which the Metric System is permissive.*

STATE.	YEAR.	POPULATION.
Great Britain...................	1871	31,817,108
United States...................	1870	38,555,983
Total.....................	....	70,373,091

IV.—In Sweden (population (1867) 4,195,681) and Norway (1867) 1,701,478, total=5,897,159) the decimal division has been adopted without as yet the metric values.

As the peoples in the second class above may be regarded as committed to the ultimate adoption of the metric system in full, we may count, as already enlisted on this side of the question, a total of about 420,000,000.

CORRECTION.—The statement, on page one hundred and seventy, that the Third Report (made in 1821) of the British Standards Commission of 1818, gives, as the weight *in vacuo* of the cubic inch of water at 62° F., 252.724 grains, while the Appendix to the Report referred to as authority, gives 252.722, is erroneous as it respects the first of these numbers. The Report (*British Parliamentary Papers*, 1821, Vol. IV.), gives 252.72; differing from the Appendix only in the omission of the third decimal figure.

The error was made in consequence of depending on an extract from the Report (which purports to be literal) given by Dr. YOUNG in his article on Weights and Measures, in the *Encyclopedia Britannica;* in which the weight in question is stated at 252.724 grs., although the Appendix, which is quoted entire in the same article, gives as above 252.722 grs.

The bill, founded on this and the preceding reports, which was introduced into Parliament in 1822, but not passed, adopted the number of the Report above, viz., 252.72 grs. The bill introduced in the following year, which was finally, with some slight modification, passed June 17, 1824, but not carried into operation until January 1, 1826, assumed the weight in question to be 252.724 grs.—the same number which Dr. YOUNG has erroneously introduced into the extract from the Report quoted by him. The act fixed the capacity of the imperial gallon by requiring that this measure should contain seventy thousand grains of distilled water at the temperature of 62° F., weighed by brass weights under the barometric pressure of thirty inches, and stating at the same time the capacity thus determined to be 277.274 cubic inches. From these data we infer the weight of the single cubic inch of water in the air, under the prescribed conditions of temperature and pressure, to have been supposed to amount to 252.458 grains; since $252.458 \times 277.274 = 70,000.039492$, exceeding the prescribed weight of the gallon of water by less than four one-hundredths of a grain.

In the bill of 1822, the capacity of the gallon was fixed at 277.276 cubic inches; corresponding to the weight of the cubic inch of water in air (under the conditions specified) as given in the Appendix to the Report, viz., 252.456 grs. Thus $277.276 \times 252.456 = 69,999.989856$ grains; which is less than 70,000 by the insignificant fraction 0.010144. Before the final passage in 1824, of the bill of 1823, the weight *in vacuo* was thrown out, and the weight in the air substituted; and it was then explicitly declared to have been "ascertained by the Commissioners appointed by his Majesty, &c., that a cubic inch of distilled water, weighed in air by brass weights, at the temperature of sixty-two degrees of FAHRENHEIT's thermometer, the barometer being at thirty inches, is equal to two hundred and fifty-two grains, and four hundred and fifty-eight thousandth parts of a grain, of which as aforesaid the imperial standard Troy pound contains five thousand seven hundred and sixty."

Dr. YOUNG makes this explanatory remark:—"The slight discordance of the numbers of the successive years depends merely upon the adoption of a standard Troy pound better authenticated than the two pound weight particularly employed by Sir GEORGE SHUCKBURGH, which [former] was finally preferred, both as representing a unit, and as being more simple in its form than the two pound weight."

Notwithstanding this, the British authorities, both governmental and scientific, constantly cite the original value assigned by KATER for the weight *in vacuo* at 62° F. of the cubic inch of water (viz. 252.722 grains); while they nevertheless adopt the corrected weight of the act of Parliament of 1824, for the weight of air (viz. 252.458 grs., deduced from the corrected weight *in vacuo* above given, viz. 252.724 grs.), though the two are inconsistent with each other. Thus, in the Sixth Appendix to the Second Report (made in 1869) of the British Standards Commission of 1868, Mr. H. W. CHISHOLM, Warden of the Standards, says that "the legal weight in a vacuum of a cubic inch of water at 62° F., authoritatively and scientifically determined by SHUCKBURGH and KATER, is 252.722 grains." But he gives also, at the same time, the legal weight of the cubic inch of water in the air, under the conditions specified in the statute, precisely as given in the statute itself, as of course he was bound to do; though this is not the value given by SHUCKBURGH and KATER, nor is it deducible from their value for the vacuum, given by Mr. CHISHOLM as the proper one, immediately before.